DYING TO CROSS

ALSO BY JORGE RAMOS

The Other Face of America

No Borders

The Latino Wave

DYING
^{TO} CROSS

The Worst Immigrant Tragedy
in American History

JORGE RAMOS

*Translated from the Spanish by
Kristina Cordero*

 An Imprint of HarperCollins*Publishers*

HarperCollins books may be purchased for educational, business, or sales promotional use. For information, please write: Special Markets Department, HarperCollins Publishers Inc., 10 East 53rd Street, New York, NY 10022.

FIRST EDITION

Designed by Daniel Lagin

Printed on acid-free paper

Library of Congress Cataloging-in-Publication Data
is available upon request.

ISBN 0-06-078944-1

05 06 07 08 09 DIX/RRD 10 9 8 7 6 5 4 3 2 1

In memory of the nineteen immigrants who died in Victoria, Texas

Marco Antonio Villaseñor (cinco years old), Mexican
Jose Antonio Villaseñor, Mexican
Serafín Rivera, Mexican
Roberto Rivera, Mexican
Héctor Ramírez, Mexican
José Luis Ramírez, Mexican
Elisendo Cabañas, Mexican
Edgar Gabriel Hernández, Mexican
Juan Carlos Castillo, Mexican
Ricardo González, Mexican
Oscar González, Mexican
Catarino González, Mexican
Juan José Morales, Mexican
Mateo Salgado, Mexican
Chelve Benítez, Mexican
Rogelio Domínguez, Mexican
José Felicito Figueroa, Honduran
José Mauricio Torres, Salvadoran
Stanley Augusto Vargas, Dominican

The happy and the powerful do not go into exile.

—Alexis de Tocqueville (1831)

CONTENTS

PREFACE

This is the story of the most shocking immigration tragedy to have occurred on American soil. Never before had so many undocumented immigrants died in one single incident. One night in May of 2003, nineteen people died of asphyxiation, dehydration and heat exposure as the result of being trapped inside a trailer truck.

Among the dead was a 5-year-old child.

On that sweltering, humid, spring evening, the victims were trapped inside the trailer, unable to open its doors as the truck lumbered down the highway between Harlingen, Texas and the city of Houston. Inside the trailer were at least seventy-three undocumented immigrants who had crossed the Mexican border in the hopes of making their way to the United States, to search for work and a better way of life. There may have been more inside; no one knows for sure.

We do know that among those accounted for, there were forty-eight people from Mexico, fifteen from Honduras, eight from El Salvador, one from Nicaragua and one from the Dominican Republic.

Not everyone paid the same amount of money to the "coyotes," the traffickers of undocumented immigrants who would ostensibly help them across the border and get them to Houston, Texas. The Mexicans, for example, paid less: geography was on their side. The Central Americans, on the other hand, had to come up with small fortunes to pay their way across the Mexican border and into the United States.

Now, of course, it makes no difference how much anyone paid. It was the worst trip of all of their lives.

The first images I saw of this incident were those that were broadcast on television, captured by a local TV news helicopter circling round and round, high above a truck that had pulled over on a South Texas highway. Initial accounts were not very clear. Several people were dead, this was certain, but nobody knew how many. How, exactly, they had died was the resounding question. I heard a variety of theories—all pure speculation.

The videotaped images that first caught my attention were literally of feet. Some were shown bare, dangling from the truck's bed. Others were clad in white socks, each sticking out from the edge of the ambulance stretchers beneath them, stark white sheets covering corpses. Then I heard a report that made my blood run cold: a child had been found among the dead. His lifeless arms had stiffened around a grown man's body. Later, they were discovered to be father and son. They took their last breath in each other's arms, never having taken a breath in the country they so desperately wanted to call home. The whole situation made me breathless, and exasperated me.

Many of the victims' families found out about what happened just as I did, through the television reports that were broadcast in the United States, Mexico and Central America. Without a doubt, it must be the cruelest way to learn of a loved one's death.

When I arrived at the scene of the tragedy in Victoria, Texas, the trailer was no longer there. Nor were there any ambulances or police officers. All I found were two barbed-wire fences adorned with flowers, a scapular, and three teddy bears—one pink, one beige, one white—in memory of Marco Antonio, the little boy who had died. The grass, yellow and dried-out from the region's blistering temperatures, was littered with crosses of all sizes— some black, some white, many of them bearing the names of the dead. The area was strewn with cards and letters that by now were just scraps of paper that had faded, made illegible from exposure to the rain, wind and sun: messages written for people who were now gone. But I could imagine what they said. *For my cousin . . . For my beloved husband . . . for my dear brother. . . .*

Men were the only ones dead. All the women survived.

Standing next to me in front of that impromptu, makeshift altar of death, were four survivors of that long night: Enrique, Alberto, José, and Israel. They stared at the flowers and the crosses around them in silence. The tribute they stood before was not for them. But it might well have been. A few more minutes inside that trailer, and who knows if they would've survived. Tears welled up and slid down Israel's face first. It didn't take much longer for Enrique and Alberto to break down as well. Turning, Enrique, Alberto, and Israel began to make their way down the same road they had taken in the early dawn hours of Wednesday, May 14, 2003. They never looked back. Not once. José, on the other hand, remained right where he stood, impassive and motionless, wrapped up in some form of silent prayer, his own belated ceremony of remembrance. I gently took him by the arm and pulled him away. The air felt thick, heavy with memories.

Suddenly, I was the one who couldn't take it any more.

"The little boy," I heard myself murmur. "The little boy."

I thought of my own son, also 5 years old. Born in the United States, he never had to cross any border illegally. For a moment, I felt myself fall apart. Tears filled my eyes, threatening to spill over. Somehow, I managed to keep them at bay. But I could feel them dancing, teetering just beneath my eyelids.

I quickly dried my eyes with the back of my left hand, turned away from the flowers, teddy bears, and crosses, and asked a member of my crew for a microphone. Work: that was the only way I would be able to keep from crying, the only way I could isolate the terrible pain that had been dredged up by the memories of those four survivors. It was the same escape valve I had used when I covered the terrorist attacks on September 11, 2001, and when I had been assigned to cover wars in foreign countries. If I focus, I can steal myself away from emotion and continue to function. If I don't, I'll be overwhelmed by the moment and I won't be able to deliver any news report.

The news crew and I were there to tape a television special about the tragedy that had happened that night in Victoria, Texas, and to interview survivors. Enrique, Alberto, José, and Israel had very bravely agreed to accompany me and my production team to the very spot on U.S. Highway 77 where the truck had pulled over. Their lives had been spared, and simultaneously changed forever. In retrospect, I now realize that when they agreed to participate, there was no way they could have known quite how excruciating it would be to relive those four agonizing hours inside that truck trailer. But it was too late to turn back at that point. Admirably, they continued cooperating, as they first stoically agreed to do.

The idea for this book grew out of the Univision television program *Viaje a la Muerte* ("The Road to Death"), and it is really nothing more than a detailed journalistic account with firsthand testimony of what happened that night and the following dawn. This book does not attempt to present an exhaustive description of each and every detail of the case—the judicial files, the official reports, and the police investigations have already taken care of that. My sole intention is to tell the story from the point of view of those who actually lived through it. Nothing more. This is their testimony to me. I owe it to the victims and the survivors to keep pure the events of their experience.

All the dialogues and statements included in this book are firmly grounded in fact: they have been culled either from my own conversations with the survivors or from official case records and news reports. Nothing has been invented. This is not fiction. It is journalism, not literature.

The facts presented here have not been modified for literary or any other kind of dramatic effect. The death of 19 human beings is shocking enough in and of itself.

Four survivors, holding back nothing, gave me a step-by-step account of what happened inside that trailer. Their tales are heartbreaking, but they are also true stories of courage and of a person's will to survive. Their stories raise questions, some very serious ones: Why didn't anyone stop the truck as it traveled down that road? Did the driver hear the immigrants' cries for help? Why didn't he activate the truck's air conditioning system earlier? What happened with the telephone call (we know of at least one) that was made from inside the trailer, seeking help? Who could have possibly thought to put so many people inside that trailer? And who the hell would have put a 5-year-old child in there?

In addition to the fact that this case set a new record for the number of immigrants dead in one single incident, it is also significant in terms of legal precedents: this was the first time that immigrant traffickers could have been sentenced to death. Nothing like this had ever happened before.

The information concerning the "coyotes" and others who have been charged in connection with this incident is based on the accounts of various survivors, statements given by several of the accused after they had been detained by law enforcement, and various police and federal law enforcement reports. It is important to stress that other than those individuals who have pled guilty to crimes in connection with this matter, these individuals have not been convicted of any wrongdoing. The extent of their involvement will have to await judicial determination.

I also hope to make it very clear that the "coyotes" involved in this incident are not the only ones responsible for these 19 deaths. Severely flawed U.S. immigration policies (under several administrations), as well as dire economic and social conditions in Mexico (also sustained over the course of various administrations), are partially to blame for what happened. Nobody's hands are entirely clean. No one's.

This case is significant because it has reopened the debate regarding illegal immigrants, an issue that the United States is under tremendous pressure to resolve. The *status quo,* the present situation, is completely untenable and unacceptable. Millions of people cannot be forced to live in fear and darkness. This is not just a humanitarian question, either; it is also a question of economics, and national security as well.

But first, I offer the story of those who died in their attempt to cross the border into the United States, and the story of those, like Enrique, Alberto, José, and Israel, who survived to tell their story.

1

WHEN THE
DOORS OPENED

I t smelled of death.

When truck driver Tyrone Williams opened the door to his trailer on the morning of Wednesday, May 14, 2003, he never would have imagined that he would find so many people inside. Or that several of them would be dead. Surprise can be such an unwelcome visitor.

As he pulled the lever and opened the door to the trailer of his eighteen-wheeler, he had to move quickly in order to avoid being crushed by the swell of humanity that spilled out, gasping for breath. Some of the bodies simply fell to the ground, motionless, not seeming to breathe at all. One glance was all it took to realize that something was very wrong. Very, very wrong.

Inside the trailer, dozens of people were strewn across its metal flooring: some were unconscious, while others merely seemed to be sleeping. Seventeen were dead, and two more would die in the hospital later that night. At that moment, however, there was no way to know exactly who had perished and who was on the brink of death. It was two in the morning, and there wasn't a soul on that rural road, just off U.S. Highway 77 in Victoria, south Texas.

There was no light inside the trailer, and there were no flashlights handy, either. The only light the panicked group had to penetrate the thick cover of night was the yellow glow of the Texan moon. The lights of a faraway gas station filtered in through one of the trailer doors, creating a thin, whitish line along the horizon. Inside, the dim shadows seemed to suggest piles of sweating flesh and broken wills. Not everyone jumped out of the trailer. Walking like zombies, some people found their way to the door of the truck and, with difficulty, lowered themselves down the two or three steps that separated them from the ground. The few people who still found themselves with a bit of strength left in them helped the others out of the truck. When the doors were opened, some had regained consciousness, and with painstaking effort dragged themselves toward the doors. Those who remained inside the trailer scarcely moved. Some were still as stone.

We will never know exactly how many people were traveling inside that trailer. If we count the nineteen who died and the fifty-four who survived (and who were then detained by the police), we know that there were at least seventy-three. Of the nineteen who died, sixteen were Mexican, and the other three were from El Salvador, Honduras, and the Dominican Republic. Of the fifty-four survivors who were identified, thirty-two were from Mexico, fourteen were from Honduras, seven were from El Salvador, and one was from Nicaragua.

But how many escaped? There may have been eighty people inside the trailer. Maybe more. Some news reports suggested that there may have been up to one hundred. We don't know. We will never know. What is very probable, however, is that some of the younger, stronger survivors managed to escape once the doors were opened. They wouldn't have been able to help much if they had stayed. They didn't really know each other as it was, and their staying would've only put them at risk. The immigrants inside that trailer had not formed strong bonds of friendship, and the majority of them were not united by family ties, either. This was not a primary concern for them, then, and if they managed to escape, they could skip out on the last installment of the coyote's fee. At the end of the day, even they wanted something for nothing.

Tuesday, May 13, was one of the hottest days in Texas that spring season in 2003. Shortly after noon, the thermometers hit 91 degrees Fahrenheit, one degree shy of the record for that date. It didn't rain at all that day, so the heat held steady throughout the night. The worst, however, was not the heat, but the humidity. The humidity and heat are so intense in that part of Texas, that it is easy to perspire through a shirt after walking a block. Clothes stick to one's body like adhesive tape. When the trailer doors were opened at that early dawn hour on May 14, the temperature had gone down a bit to 74 degrees Fahrenheit. But the relative humidity, at 93 percent made it feel like a hot rainstorm.

The weather conditions turned that trailer into a sauna. The high daytime temperatures, the humidity, and the heat generated by so many dozens of bodies pressed against each other turned the trailer into a deathtrap. There is no way to know exactly how high the temperatures rose inside the trailer, but the

Associated Press, citing local authorities, suggested that it may have actually hit 173 degrees Fahrenheit. There is of course no way of knowing for sure.

The trailer was hermetically sealed shut, for a very simple, commercial reason. This type of trailer often transports perishable goods: vegetables, fruits, meat, and other food items. The less air that enters the inside of the trailer, the longer the merchandise remains intact, and the farther these goods can be transported. These trailers are not outfitted to transport human beings.

The walls, the ceiling, and the floors were all lined with one layer of aluminum and then another layer of insulation. This inner structure ensures that the temperature remains constant inside the trailer, even if there are shifts in the outside temperature. Even if the immigrants had been able to cut through these two layers, they would have then found themselves facing the steel shell on the trailer's exterior. It was impossible for the trailer to be opened from the inside. Once inside, there was no way out.

This type of trailer is equipped with an air conditioning system, which can be used to keep merchandise refrigerated if necessary. Yet for some inexplicable reason, the air conditioning system was not turned on that Tuesday night, until it was too late. A number of the immigrants noted that the air conditioning began working just a few moments before the doors opened. Once it had, it was useless. Too late.

The trailer had covered some one hundred and sixty miles since departing from Harlingen, close to the Mexican border, and still had about one hundred miles left to go, heading north-east toward Houston, its final destination. The driver seemed to have been in no rush at all, driving at an average speed of fifty miles an hour, significantly lower than the speed limit. The explanation for this is very simple: he was no doubt being very prudent so as to avoid getting pulled over for speeding. The driver appeared to take every precaution to avoid inspection of his truck by the highway police.

The immigrants who survived estimate that they spent four hours inside the trailer, unable to get out. If for some reason the driver, Tyrone Williams, had decided not to stop in Victoria, it is very likely that all the people traveling inside would have perished. At the speed Williams was driving, it would have taken him at least two more hours to reach Houston.

By the time the authorities arrived, almost an hour had gone by since the truck door had been opened and the first immigrants made their way toward fresh air. During this hour, some of the bodies had been taken out of the trailer. The first group of agents from the Victoria County Sheriff's office found thirteen bodies inside the trailer and four more lying on the ground outside.

Henry García Castillo, deputy sheriff of Victoria County, was one of the first to arrive, at three in the morning. He remained at the site for seventeen hours.

"The first thing you investigate is whether there is anyone alive and if you can help them in any way," he said, in rusty Spanish, to a Spanish-language TV news reporter. "After that, you have to look for evidence. That's when you start the rest of the investigation."

What most shocked Sheriff García Castillo when he arrived

was the sight of a little boy, dead, lying next to his father. "[The boy] was inside the trailer," the sheriff explained. "And from what I understand the man lying next to him was his father. It broke my heart."

The morning and afternoon of Tuesday, May 14, would remain forever etched in Sheriff García Castillo's memory. He was part of the team that brought the survivors who were ill—including two in very critical condition—to the hospital, along with those who were suffering from heat exposure and dehydration to the Victoria community center.

"There was a young girl in the community center, it was her birthday," the sheriff recalled. "And the people went and got her a cake so that she could celebrate her birthday, so that there would be something that day that wasn't so sad."

I wondered how this little girl would later celebrate her birthday in the years to come.

It was already dawn by the time the news reporters arrived at the spot where the truck had pulled over. The paramedics that had arrived on the scene were still treating some of the victims, but the photographers and cameramen couldn't resist capturing some images of the victims' bodies, covered in white and gray blankets, their feet sticking out. Some still had their socks on.

The images filmed from the TV news choppers high in the air showed the long trailer next to a barbed-wire fence, surrounded by dozens of patrol cars with their red and blue lights flashing overhead. Several ambulances and the satellite trucks from the local TV stations also lined the side of the road. Someone had brought wheelbarrows to the scene. They were piled high with the bodies of the dead, each body waiting to be taken to the morgue. Long yellow barriers blocked the area marked off-limits to everyone but the police investigation team, and reporters and cameramen interviewed, over and over again, the same local authorities who still had very little information about what had really transpired that night.

The Wednesday morning sun rose, and a gloomy feeling of frustration, anger, and shock weighed heavy in the air. Who were the people responsible for these deaths? Who let them die? Who killed their dreams? How did the survivors make it out alive? Where was the front part of the trailer? Who was that little boy?

The answers would soon make themselves known.

2

THEY CAME FROM FAR AWAY

At least seventy-three people had made the necessary arrangements to illegally cross the U.S.–Mexican border and then travel from the city of Harlingen, Texas, to Houston. But none of them knew what they were really getting into.

This was not Enrique Ortega's first trip to the United States. In 1991, at age 15, he had entered the country illegally and settled in Houston, where he scraped by as a restaurant worker, first as a cleaning assistant and eventually as a chef. Not bad for a kid who had been earning the equivalent of $3.50 a week as a poor farm laborer in Tecomatlán, Puebla. Enrique had also been a factory worker and a sales clerk in Puebla, but there he had never earned more than $50 a week. Enrique could earn that much in a single day in Houston. Who could possibly tell Enrique he was wrong to try and leave Mexico?

Enrique is the eldest son in a family of nine brothers and sisters.

"I come from a poor family. There are no good job opportunities for us in Mexico," he said, trying to explain the reasons that drove him to emigrate north. "And here, [in the United States] we can get work, we can support ourselves and even help our family out a little [in Mexico]."

Like many immigrants in the United States, Enrique settled down, fell in love, got married, and had three daughters. But in the year 2000, while driving his car, he was stopped by the police for committing a serious traffic violation. The police, in turn, handed him over to the Immigration and Naturalization Service, and he was deported to Mexico. Before he was deported, a judge warned him that if he was arrested again in the United States, he would be put in jail for one or even two years.

Enrique waited for three years in Mexico so that his wife, a U.S. citizen, could file the papers he needed for his residence permit. But Enrique was exasperated by the slow bureaucratic process, worried about being separated so long from his wife, and desperate to see his daughters, so he decided to return to the United States in spite of the judge's warning and the threat of ending up in jail.

Enrique took a bus from Puebla to the border town of Matamoros, in the Mexican state of Tamaulipas. There, he telephoned someone whom he identified as Víctor Rodríguez, a man he thought might be able to help him cross the Texas border and get him all the way up to Houston.

"Two thousand dollars from Matamoros to Houston," said the coyote, or *pollero,* as some people call them. Enrique accepted the deal.

Late at night on Monday, May 12, 2003, one of the coyote's contacts went to find Enrique at his hotel. That very night they traveled from Matamoros to Brownsville, Texas, without too much difficulty. According to Enrique, that was where he saw Víctor Rodríguez for the first time and paid him the initial sum of $1,000.

Enrique said that Víctor Rodríguez's wife then picked up the telephone and called a woman named Karla Chavez.

"Karla, do you have space for two more?" she asked. Karla, apparently, said yes. She would hold two spaces for Enrique and an immigrant from the Dominican Republic. To confirm the reservations, however, Víctor Rodríguez's wife said they had to go to a parking lot in front of a restaurant near Harlingen and give the money to Karla. That was where Enrique says he saw Víctor Rodríguez pay Karla for the transfer of two people from Harlingen to Houston.

Enrique would be expected to hand over the balance, another $1,000, when he arrived at Houston. After making his first payment, Enrique was taken to a house in Harlingen. From there they would leave for Houston on the night of Tuesday, May 13.

Those who are unfamiliar with the methods of the U.S. Immigration Service may think that the risk of being stopped and deported ends once immigrants get across the border. But this is not the case. Sometimes, crossing the border itself is in fact the easiest part of the trip, and the hard part is actually the journey from the border zone to a destination city such as Houston, Dallas, Phoenix, or Los Angeles. Once in those cities, people have a much easier time moving on to other parts of the United States. In a sense, there exists a kind of second border within the United States, an area known only to the immigration agents, the undocumented immigrants, and the coyotes that guide

them through this terrain. This second border is far more difficult to cross than the physical, legal border separating the
United States and Mexico.

Enrique was dressed in a short-sleeved summer shirt. When
Víctor Rodríguez saw that he had nothing else to cover him, he
said, "I'm going to get you a sweater or something, because
you're going to be cold." The sweater never materialized. Enrique thinks that Víctor Rodríguez forgot about the sweater because he was drinking. In any event, Enrique felt encouraged by
the thought that he might be cold; that meant he would be traveling to Houston in some kind of air-conditioned vehicle.

Enrique had been told that he would be traveling in a trailer.
But he never imagined that he would be traveling with dozens of
other people in a truck container. Enrique thought that he
would be traveling in "the cab part . . . not in the trailer." They
told him he would be traveling with five women from Honduras
and three other men. That, however, wasn't the case.

The night of March 13 turned out, as superstitious types
might have predicted, to be lined with bad luck. From the house
in Harlingen, Enrique and dozens of others were taken to the
outskirts of the city. Shortly thereafter, a tractor-trailer pulled
up, and they were told to get in. Enrique had his doubts.

Suddenly, a movie flashed through his head: *El carro de la
muerte,* the Death Car. He had seen it when he was 10 years old.
It was about a group of undocumented immigrants who got into
a water delivery truck and died.

"I thought back to that movie, and I had a feeling the same
thing was going to happen to us," Enrique said. "Everyone had
already jumped into the trailer except me. I was the last one to

climb in . . . I had a feeling something was going to happen. But I also thought, 'If I don't get into that trailer, [the Immigration Service agents] are going to catch me and throw me in jail, and then I'll lose the thousand dollars I gave the coyote.'"

Enrique hesitated for a few seconds more, but eventually climbed aboard, a last-minute decision that would very nearly cost him his life. It was approximately 10 p.m. The weather reports predicted a hot night in south Texas.

At the back of the trailer, along with dozens of other people, was Alberto Aranda Amaro, a 23-year-old man who was traveling to the United States for the third time. Alberto had two dreams: to become an engineer and to build a house for his mother, who suffered from diabetes. So many times before, Alberto had looked out at the houses in his neighborhood just outside of Mexico City, dreaming that one day his mother would be able to live in the biggest and prettiest house of them all.

Alberto's intention to become an engineer was thwarted by his family's financial situation. He has two sisters, one younger and one older. As the only young man in the house, he had to get out and look for work. Alberto wanted to build houses but he hadn't realized how difficult it would be, and how little he could earn unless he was an architect, engineer, or contractor.

One of his uncles offered him a job as a bricklayer for $20 a week. With few other options, Alberto accepted the offer and remained on the job for two years. But one day, as he laid the bricks and spread them with cement, he decided it was time to try for a better life in the United States. It was 1997.

"I told my mother that one day I would build her a house," he recalls. " 'I have two hands, Mamá, and I'm young. Maybe I didn't make it through school, but I'm going to show you I can do it.' " That's what he told her.

And that is what Alberto did. He took a bus to Nuevo Laredo, in the state of Tamaulipas, at the border between Mexico and the U.S. There, he and four other men decided to enter the United States without the aid of a coyote. If they managed to pull it off, they would each save $1,000. One of Alberto's friends, moreover, claimed that he knew of a route that was quick and safe. It didn't turn out quite as they planned.

After crossing the Rio Bravo (or Rio Grande, as it is known in the U.S.) without too many problems, they got lost and had to walk for almost five days until they finally reached Crystal City, Texas. From there, for the sum of $300, a *pollero* took them to Houston. The risk had paid off.

During this first time living in the United States, he found his way to North Carolina, where he began working in construction. He left when he found a job at at a paint company. He worked there for two years and then returned to Mexico.

In the year 2000, Alberto crossed the border for the second time, via the same route he had taken on his first trip to the U.S. He worked for the same paint company in North Carolina for another year and a half, returning to Mexico in mid-2001. Nevertheless, every time he returned to Mexico he would inevitably grow exasperated at his inability to earn enough money to build an addition to his mother's house. He had made a few changes here and there: the roof, for example, no longer leaked thanks to his handiwork.

"At least we could sleep easier after that," he acknowledged. But the house still needed more work. A lot more work.

For this reason, Alberto decided to go back to the United States for the third time in 2003. Instead of crossing over from Nuevo Laredo—where he feared that border security had grown tighter—he decided to make the trip from the city of Reynosa.

This time, the journey was much more expensive and far more difficult than the other two trips he had made to the U.S.

Alberto could no longer risk crossing the border without the help of a coyote, as he and his friends had done six years earlier. Times had changed; there were more immigration agents than ever stationed along the border. The new Bureau of Customs and Border Protection, which is part of the Homeland Security Department, had replaced the Immigration and Naturalization Service and by 2003 had some 10,000 agents on its payroll. This personnel increase was not just a direct response to the terrorist acts of September 11, 2001, in Washington, New York, and Pennsylvania. It represented, in fact, the implementation of a new immigration policy that had taken effect a decade earlier, a program that focused on improving security along the border. The result of this change was that immigrants were now forced to take more distant, dangerous routes and to rely on coyotes to ensure their crossing.

Once Alberto got to Reynosa, Tamaulipas, some acquaintances from North Carolina connected him to a coyote. The deal was simple: $2,000. One thousand to be paid in Harlingen and the rest when they arrived at Houston. The days when you could get up to Houston for $300 were a thing of the past, a very distant past.

Alberto's coyote, sometime around two in the morning, led him by foot from Reynosa to McAllen, Texas. Two other men and a woman joined Alberto and the coyote a bit farther down the road. Within three hours they had crossed the river and made it to McAllen, Texas. The coyote then made a call from a cell phone, and after a few minutes, a pickup truck pulled up and brought them to a house. Shortly thereafter that same morning, they took Alberto to another house in Harlingen. There, he paid the first thousand dollars and was made to wait for the next six days. Then he was delivered to another house, where he waited two more days. During this long limbo period, the various people involved in this immigrant trafficking network told Alberto that he and three other people would be traveling very comfortably to Houston in the driver's cab of a tractor-trailer. They were told that the trip would last four hours. Nobody ever told him that he would have to travel in the truck trailer, without air-conditioning, in the company of dozens of strangers.

After waiting around for eight days, Alberto's heart skipped a beat when he heard the news: finally, on the night of May 13th, they would be leaving for Houston. At around eight in the evening, the immigrants were taken out to a desolate spot somewhere outside McAllen, Texas. At 10 p.m., the trailer arrived. This is how Alberto remembers the moment he saw the trailer pull up:

"They never told me that I would have to go in the trailer part of the truck or that I would have to travel with so many other people. If they had, I might not have agreed to any of it, you know? When I saw how many of us were up there on the hill, I figured that more trucks were coming. But when they opened the door and said, *'vámonos,'* and everyone began running toward the truck . . . that was when I really started to get scared."

Still, Alberto got in. He figured the ride would be short, that he and the other immigrants would travel together until passing the inspection booth, after which point they would be split up into various pickup trucks. Alberto sat down in the middle of the trailer. Others, after a while, began to stand up. But Alberto made an effort not to let his desperation get the best of him, because he knew he was better off down by the floor. He forced himself to think about his future, his plans to build his mother a better house.

"What if I can't take it?" Alberto wondered when he saw so many people packed inside the container. "Ever since I was a little boy, I've always been nervous. Everything scares me. My hands start to sweat." Little did he know that in a few short minutes he would be sweating from every pore on his body, not just his hands.

José Reyes Arellano, 47 years old, had a tough time climbing into the truck. And not just because he was one of the oldest undocumented immigrants in the truck that night, either. He was also one of the weakest. José suffered from diabetes and asthma. But even that didn't stop him.

José, originally from the town of Pozos in the Mexican state of Guanajuato, had a debt to pay off, and the kind of jobs available to him in Mexico would never have earned him enough money to pay what he owed and save a little afterwards. But José had other concerns, too.

When José's granddaughter was born two-and-a-half-months prematurely, special medical attention had been required in order to save her life. José had had to cover these medical expenses. At the same time, his 14-year-old daughter had gone to the doctor on numerous occasions to get treatment for some severe leg pain and also for her vision problems: she couldn't see anything beyond three or four meters away. José took her to an ophthalmologist, who helped improve her vision somewhat. But José hadn't paid his medical bills, and he had no idea how to come up with the money. Needless to say, neither José nor his family had health insurance.

José described himself as a man of "limited resources." This situation, complicated further by the medical expenses incurred by his granddaughter and daughter, was what drove him to go to the United States. But José didn't travel alone. With him were his brother-in-law Hector Ramírez and three other relatives: Roberto, Roberto's brother Serafin, and José's nephew Israel. Of those five, only two would survive the trip.

Israel Rivera Sánchez, José's nephew was 23 years old when his uncles suggested he join them on their journey to the U.S. It wasn't tough to convince him. Some time earlier, Israel had been talking on the phone with his older brother—Israel is the second-oldest son in a family of eight—who lived in Florida. Israel already had some money saved, but not enough.

"I said to him, 'If you'll help me, then yeah, I'll go.' " That is what Israel remembers telling his brother. And that was how he decided to make the trip.

Israel seemed timid when I first met him. Slowly but surely, however, he grew more and more comfortable and began to talk at length, without pausing, stumbling on his words as they tumbled out of his mouth. There wasn't much to keep Israel in Pozos, Guanajuato. Despite the fact that Guanajuato is the home state of Mexico's president, Vicente Fox, job opportunities for Israel and his family did not improve following Fox's election in 2000.

Fox had wanted to transform Pozos into a tourist destination. But the town's few hotels and restaurants were barely enough to employ even a minimal fraction of the three thousand people living there. A century ago, Mineral de Pozos was one of the country's most important commercial centers, and the people of Pozos became some of the most important players in the Mexican mining industry at the end of the 19th century. About five hundred different companies—both domestic and foreign—took full advantage of the local mines, over three hundred in total. Pozos boasted a population of 80,000, and its economic impact was such that the former dictator, Porfirio Díaz, declared the town a municipality and baptized it with a variation of his very own name.

Pozos was founded in 1576. The Jesuits, taking advantage of the very primitive technique used by the local indigenous people, were the first to exploit the minerals in the area. In 1844, with the discovery of mercury, the mining industry went through a boom period, and soon after, massive gold and silver deposits were also found in the area. A number of American companies began doing business in Pozos, but the Mexican Revolution in 1910 effectively halted the mining work, and many of the mines were flooded. The Cristero War, fifteen years later, sealed the doom of what had been one of the most prosperous towns of the Americas during the 19th century.

At the dawn of the 21st century, Pozos was busy trying to recapture the glory that it had enjoyed during the days under Porfirio Díaz. Various old hotels that were rundown but still quite imposing were undergoing major renovations, and Pozos was being touted as an ideal tourist destination only forty-five minutes by car from the more-renowned town of San Miguel de Allende. Pozos's main attraction was its collection of old mining-era haciendas: Cinco Señores, Angustias, San Baldomero, and Santa Brígida. A stroll down the cobblestone streets of Pozos was like taking a walk through time from pre-Columbian Mexico into the Spanish colonial period and finally culminating in the modern era, at a pair of new hotels that advertised their rooms on the Internet. A double room in the Hotel Casa Montana, to give one example, could cost up to $100 a night. A fortune for any Pozos native.

That was twice the amount of money Israel earned for a week's worth of work. But because he didn't speak a single word of English, there was no way he could get a job at the new hotels that catered to tourists from the U.S. and Europe.

Ironically, Israel spent his time in Pozos digging *pozos*—wells. But that was only when there was work to be had. Sometimes he went for an entire month without getting hired for a job.

"There wasn't enough work. That's why I came here," he said. He had figured that he would cross the Texas border with his relatives and then move on to Florida, where he could work with his brother. But he never would have imagined that the complications would begin even before he left Mexican soil.

Five people—Israel, his friend Hector, and his uncles José, Roberto, and Serafin—took a bus from Pozos to Celaya, also in the state of Guanajuato, very early on the morning of Monday, May 5, 2003. Later on, they continued down the road toward Reynosa, Tamaulipas, close to the Texas border. But their troubles began as soon as they stepped off the bus.

"You people, get over here," a policeman called out to them as they arrived at the bus terminal. "Where are you from?" Guanajuato, they replied. "Do you have family here?" the policeman asked, in an insistent tone.

"No," José innocently replied. "We're here to cross the border, to get to the other side."

No doubt the policeman's eyes sparkled at that moment. Nobody crosses the border without carrying cash. Everyone knows that coyotes don't accept Visa or American Express.

"All right then," the policeman said. "Well you're going to have to cooperate with us first." They knew what that meant: the policeman wanted a little payment. A kickback. At first they refused, but the policeman then threatened them, saying that if they didn't give him something, he would call in the "higher authorities." They never found out who those "higher authorities" were, but they did know that if more policemen got involved, they might lose all the money they had with them. The five men placed their wallets on a table, just as the policeman had told them to, and left the room. When they returned, the police had taken $400.

Despite this setback, they had in fact managed to hide the

majority of their cash from the policeman, which meant they would still be able to get to the other side. As they walked out of the Reynosa bus station, they crossed the street and Hector—who was José's brother-in-law as well as a family friend—placed a telephone call to a contact person who would ostensibly help them cross the border into Texas. This person told them to go to the Hotel Cancún and then call him back from there. Hector did exactly that.

Twenty minutes later, a white car arrived at the hotel, picked them up, and brought them to a house in Reynosa. There, they were informed that at two in the morning they would depart for the United States. The fee was $1,800 per person: half upon arrival at Harlingen, half in Houston. They didn't have all the cash, but they accepted the deal anyway. Israel's brother, in Florida, had promised to wire them the rest of the money once they arrived safe and sound.

That dawn, they walked for about an hour until they reached the Rio Bravo. There, tires and inner tubes were distributed so the travelers would be able to float as they crossed the river. They removed their clothes and shoes. There were no plastic bags around to protect their clothes, so they gathered up their things and bundled them up inside a shirt. They would get wet, but at least they wouldn't lose them. All five of them were nervous; none of them felt very confident about their swimming abilities. Without a doubt, though, Israel was by far the most nervous of them all.

"There were waves coming off the river, and I was scared," Israel recalled. He didn't know how to swim. "If I had fallen in, I would have died, or . . . well, I would have done whatever I could [to save myself], but the truth, man, no way." The coyotes then dragged them across the river, with two people clinging on to each inner tube. All their belongings got soaked, but they didn't care—they were on the other side of the border now.

They kept walking until they reached a hill in the outskirts of Hidalgo, Texas, where a van picked them up and took them to Harlingen. That was where one of the coyotes—that is, one of an ever-increasing number of people involved in this operation—asked them for the first half of the fee they had agreed to pay. They then were made to wait at a house in Harlingen for approximately a week, during which time more and more immigrants like them kept on arriving, with the same hope of getting safely to Houston.

The night of Tuesday, March 13, four people came by the house, herded everyone into vans, and brought the immigrants to another house, also in Harlingen, where all the people who would later board the truck were waiting. That was when Israel began to sense that something was very wrong.

"Damn, it's hot in here," Israel said. Then he heard another immigrant say that eighty-three people were scheduled to travel that night. We don't know whether he actually counted them himself or if he heard this estimate from someone else. But it is one of the few indications we have as to the total number of people that may have been inside the truck container that night.

On a hill outside of Harlingen, the coyote team began taking groups of people out of the vans. The last group arrived just as the truck pulled up. As the large vehicle came to a halt, someone shouted:

"Hey, come on over here! Get in!" Serafin and Roberto were

among the first to board, with José and Hector following closely behind. Israel, shaking all the while, inched toward the door of the container. "If it's got air conditioning, we'll get to Houston all right," he said to himself at that moment. "But if not, it's going to be tough. And then what will we do?" Before he could answer the question, one of his uncles grabbed his hand and pulled him up into the truck trailer.

Inside, everyone was silent. The last thing they wanted to do was make noise and draw the attention of another driver on the road or, even worse, the police. Israel was still worried about the lack of air conditioning.

"All right, man, let's see if they put the A/C on," Israel said to his uncle, already inside the truck.

"Get in!" they heard the coyotes shout again.

The doors finally shut, and the driver started the engine. When next the doors opened, nothing would ever be the same for three of the five immigrants from Pozos, Mexico. It would be the last trip of their lives.

It had been two weeks since driver Tyrone Williams had gotten a good rest. On Thursday, May 1, 2003, he had driven from New York to Edinburg, Texas, to pick up a shipment of melons for Bagley Produce. While waiting outside the fruit warehouses, two Hispanic men driving a green Ford Explorer had approached him, identifying themselves as "Joe" and "Abel." From what Williams could tell, "Joe" was about 29 or 30 years old, and from Central America. He was stocky, slightly overweight, with dark skin, short hair, and a goatee. Abel had dark skin, too, and he looked like the kind of man who always ate a little more than he should. For the record, Abel was eventually identified as Abelardo Flores, and Joe was, in fact, an alias for Alfredo "Freddy" Giovanni García, Karla Chavez's 33-year-old boyfriend.

During their conversation, Joe and Abel asked Williams how much money he stood to make shipping those melons up north.

"Twenty-eight hundred dollars," Williams answered. Right there and then, Joe and Abel offered Williams $3,500 to take their "cousin" to Houston. Don't worry about the immigration agents or the surveillance booth on the highway, they said.

"The checkpoint is taken care of," they told him.

As they talked, a tan Dodge Caravan pulled up next to Joe and Abel. Williams, later on, would tell the police that a tall, attractive Honduran woman had gotten out of the van. Who was she? Williams asked the two men. "The *jefa,* the boss lady," they told Williams, who did not speak Spanish. Joe and Abel went on to explain that she was the person who had the connection with the immigration surveillance checkpoint. Williams never found out her name.

Despite Joe and Abel's very tempting offer, Williams did not accept, at least not right then. He did, however, jot down their cell phone numbers in case he changed his mind. After all, he was planning to be back in Texas the following week. And that was how it happened.

Williams drove his shipment of melons from Texas up to the Market Basket in Massachusetts on Thursday, May 8. But the shipment was refused, and so he had had to drive to Connecticut and deliver it to a store there. With very little time to rest he then turned around and headed back for Texas on Friday, May 9. This time around, he was carrying a load of lactate milk for HEB stores in San Antonio. He couldn't drive empty. To keep himself in business, he had to keep his three trucks busy at all times.

Tyrone Mapletoft Williams, a Jamaican immigrant born in 1971, had permanent residency in the United States. By dint of hard work and effort, his business had grown bigger: he now had three tractor-trailers and an employee to help drive them. Not bad for someone who had started out at the bottom of the pecking order as a driver. His company, Tyrone II Transport, was his pride and joy. He did the driving, and from their house in Schenectady, New York, his wife Karen coordinated the various jobs that came in from all over the country. But things hadn't been going so well lately. One of his trucks had broken down in the Rio Grande valley, and he needed money to fix it.

Williams took his time making this trip to Texas, his second in ten days. On a pit-stop in Cleveland, Williams arranged for a friend, Fatima Holloway, to accompany him. Once he got back on the road to Texas, Williams took out the slip of paper where he had jotted down the cell phone numbers that Joe and Abel had given him, and he called them to tell them that he could be in Harlingen on Tuesday, May 13, after dropping off a shipment of lactate milk in San Antonio. From that moment on, Joe and Abel maintained constant contact with Williams.

As Williams unloaded his delivery at the HEB warehouse in San Antonio, he received a call from Joe, asking him where he was.

"In San Antonio," he replied, and then asked them to find

him a hotel room near Harlingen. Once he finished unloading his shipment, Tyrone Williams and Fatima Holloway took Highway 35 south toward Laredo; later on they would drive east along the border until they arrived at Harlingen. Before reaching Laredo, Williams noticed that on the other side of the highway, a group of undocumented immigrants were stepping down from a truck at one of the Immigration Service's highway checkpoints.

"That's what we will be doing," Tyrone told Fatima.

As soon as he reached Harlingen, Williams got in touch with Joe, who appeared shortly thereafter in the same Ford Explorer with tinted windows that Williams had seen him driving about two weeks earlier. Williams followed Joe to the Motel Horizon, and Joe went into the office to check them in. Shortly afterward he emerged with the keys to room 135, and the three of them went in. After looking over the room to make sure it was in decent condition, Joe left, promising to return half an hour later.

At almost 10 p.m., after a brief rest, Williams and Holloway went back to the truck and followed a car to the outskirts of Harlingen, a rural area shrouded in darkness. Neither of them got out of the truck. Williams did, however, feel the slight shifts in the truck bed as the immigrants climbed into the container. He could also hear them talking.

He never knew the total number of people who got into his truck that night. Joe and his friend Abel had told him he would be transporting sixteen people, no more, no less. After closing the doors, they told Williams to take Interstate 77 toward Houston. He did as instructed.

3
THE COYOTES

Some people call them coyotes, and others call them *polleros,* but Karla Chavez calls them *dueños de la gente*—people owners. According to the legal documents used in the case against the fourteen people suspected of participating in the plan to transport undocumented immigrants in a tractor-trailer, the main coordinator seemed to be Karla Chavez, a native of Honduras. She, however, always denied this.

Karla's problem was that several undocumented immigrants had told the agents from the new immigration service, the U.S. Immigration Customs and Enforcement (ICE), that she had coordinated their illegal entry into the United States and that other coyotes or immigrant traffickers had gone to her to reserve spaces in the trailer that was supposed to drive dozens of immigrants up to Houston.

Karla had never been arrested in the United States before, but because of her line of work she was in constant danger of ending up in jail. And she, in some way, seemed to understand this. According to Karla's statements to investigators, the father of her three children, Arturo Maldonado, nicknamed "El Morro" (and alternately identified as Heriberto Flores), was in a prison in Texas at the time, and she had had to pay $3,000 in September of 2002 to cover some of his legal expenses.

The New York Times reported that Maldonado was from the Mexican state of Guerrero, and that on three separate occasions (1989, 1995, and 2001) he had been deported to Mexico after having participated in immigrant trafficking operations. After his last arrest, which took place during a raid carried out by the immigration service, Maldonado was placed in the Cameron County jail in Brownsville, Texas, and in April of 2003 he was sentenced to seventy months in prison. With "El Morro" behind bars, Karla needed to find some way to earn a living. She did just that.

Karla was not married to Arturo Maldonado, nor did she feel particularly loyal to him. In fact, it is entirely possible that she had had some kind of dispute with the father of her three children before he was arrested. She even admits to having maintained a romantic relationship with Freddy (Alfredo Giovanni García) while "El Morro" completed his prison sentence.

Karla had met Freddy in a bar. According to Karla, Freddy worked with Abel Flores, another of the participants in the plan to transport undocumented immigrants from the southern border of Texas up to Houston that fateful May 13.

Karla and Abel seemed to have a good working relationship, one that grew even stronger when Abel moved from Dallas and settled in Harlingen. Abel had originally moved to Harlingen because of a good job prospect, but when the job did not work out, Abel found himself stuck in Harlingen with no work. That, apparently, was what motivated him to join forces with Karla in their immigrant trafficking enterprise.

The division of labor appeared very clear: Karla had the contacts to find undocumented immigrants who wanted to travel to Houston, Abel figured out how to transport them, and Freddy's job was to help Abel and to keep Karla company. One could say they were not brought together by love, but by the trafficking of undocumented immigrants. The evidence indicates that Karla and Abel were the managers of the operation. They had worked together for at least two months to transport undocumented immigrants within the United States, with great success.

There is no question that Karla had many debts of her own to pay off. Twenty-five years old, she had three children to take care of, all on her own, plus she had to help out her family back in Honduras and contribute to Maldonado's legal bills. Karla Chavez had two restaurants on Rangerville Road in Harlingen, Texas. Though she didn't have the money to buy the restaurants outright, she did have enough to rent the space. Her real goal, however, was to buy the house in Harlingen she lived in, but she first had to finish making the payments to its owner.

Based on the legal documents pertaining to the case, everything seems to indicate that Karla Chavez's restaurant business met neither her basic needs nor her financial ambitions. Without a doubt, transporting undocumented immigrants was a far riskier business, but it was also a far more lucrative one.

The "shipment" that was scheduled to depart at dawn on May 13, 2003, included a number of people from El Salvador, and Karla had personally taken care of coordinating the contacts that would facilitate their arrival in the United States. Traveling illegally from El Salvador to the U.S. is far more expensive than traveling legally. Sometimes, for example, an economy-class plane ticket from San Salvador to Miami or Los Angeles sells for less than $500. The same trip, taken illegally, can cost up to ten times as much.

Fifty-five hundred dollars was what Ana Márquez Aguiluz was told she would have to pay Karla Chavez if she wanted to travel from El Salvador to Houston, Texas, according to a statement she made to an ICE agent. Ana's father, a Salvadoran resident, had been the first person to make contact with Karla.

The story was more or less the same with Carmen Díaz Márquez's uncle. Through Karla, he was the one who had arranged his niece Carmen's journey to the southern border of the United States, for the same sum of money that Ana Márquez Aguiluz had paid: $5,500. Both Carmen and Ana would first have to cross the Guatemalan and Mexican borders; once in Mexico they would go to the port of Veracruz on the Gulf of Mexico, and there they would await further instructions.

José Martínez Zuñiga's brother knew of a woman in Texas named Karla who helped people cross the border. José's brother had given him two telephone numbers for Karla, and when he arrived at Veracruz, José called her from a hotel in the city. Karla then told him that she would send someone to pick him up and coordinate his transfer to the United States. Soon after their conversation, a woman came to José's hotel and informed him that she had been sent by Karla, and that she had purchased bus tickets that would get him, Carmen, and Ana to Reynosa, into the Mexican state of Tamaulipas. And that was exactly what happened.

The three Salvadorans endured the long ride from Veracruz to the northern border of Mexico, and once they reached the bus station in Reynosa, they were led across the border— illegally, of course—and into the United States. Their first stop was somewhere near Rio Grande City, Texas.

Finally, they had reached the United States. But they were not out of danger yet. Traveling from the southern border zone up to Houston was often far more perilous than the border crossing itself. Immigration service agents kept a constant watch over all the roads heading north from the border, and undocumented immigrants had no other choice but to hide in vehicles to avoid getting caught.

After crossing the border without too much trouble, José, Carmen, and Ana were taken to various safe houses in Harlingen, Texas. At one of these houses they say they met Karla. Right away, Karla began asking them to call their family members, instructing them to wire their payments to her, the same amounts they had agreed upon before setting off. Only after the payments were made would they be able to continue on to Houston. Karla didn't want any mistakes or delays, and for that reason she herself telephoned Santos (one of Carmen's relatives in the United States), José's brother, and Ana's sister, who was living in Washington, D.C. Over the phone, Karla instructed each of them to wire $1,500 to a Western Union office in San Benito, Texas, in the name of another woman.

While the documents pertaining to the case suggest that Karla was in charge of procuring payments from at least some of the immigrants (such as the three Salvadorans), she was not in charge of finding the vehicle in which they would travel. That was apparently Abel Flores's job.

This was not Karla's first time participating in an immigrant trafficking operation. The first time around, Karla transported twelve immigrants, and by her second job that number had jumped to thirty. She couldn't remember how many immigrants she had taken across the border in the third operation, but the fourth operation involved twenty.

This trip, her fifth time trafficking immigrants, was her biggest job yet. However, she had not been responsible for finding all the immigrants who rode in that truck, as the agents of the newly formed immigration service later acknowledged. Various *polleros* had brought their own *pollitos*—little chickens. In her statements to investigators, Karla identified each of these *polleros* and their roles, although none of them have been charged in connection with the May 13 operation:

Gabriel Chavez sent fourteen undocumented immigrants to the truck. The weekend before the trip, Gabriel had telephoned Karla to tell her how many people would be traveling. Karla knew Gabriel well; in the past they had worked together on at least three other immigrant trafficking operations. This time, though, Gabriel wanted to do things differently. In the past, Gabriel had gotten the immigrants through by having them walk around the immigration service's inspection checkpoint. This time he was going to take the risk of sending them inside the truck.

Rafa "La Canica" told Karla he was planning to put six immigrants on the truck. But then he phoned her to say that he might have eight. Karla knew "La Canica" well, because he also lived in the area of Harlingen and San Benito, Texas.

"El Caballo" had contacted Karla to let her know that he had three immigrants who needed to get up to Houston, and that he wanted to reserve spaces for them on the truck.

Ricardo Uresti, from Harlingen, wanted to send five immigrants on this trip, including a 5-year-old child, Marco Antonio, and his father, who were traveling from Mexico City. Karla knew that they "belonged" to Uresti, who worked closely with someone named "Tavo" Torres in the business of immigrant trafficking. Nobody, it seems, ever suggested that the boy was too young to endure the trip inside the truck container.

Octavio "Tavo" Torres had twenty immigrants for the trailer ride. On Friday, May 9, 2003, "Tavo" had phoned Karla asking her to reserve spaces for them.

Salvador Ortega, whose nickname was "El Chavo," had also sent some immigrants in the truck, but Karla didn't know exactly how many.

Norma Gonzalez would be sending two immigrants: one was the brother of an employee who worked at her restaurant, and the other was the granddaughter of a woman who had contacted her by telephone. Norma, 43 years old, was the manager of a restaurant in Houston. After twenty-two years in the United States, she was finally achieving the American dream. She had become a U.S. citizen, her three sons were also U.S. citizens, and she even owned her own restaurant. Norma, a short woman with dyed blond hair who weighed no more than 130 pounds, didn't seem to live badly at all. She drove three different cars: a red Chevy minivan, a blue Chevy van, and a Dodge truck.

The employee at her restaurant, who requested anonymity when speaking to authorities, testified that she knew that Norma helped immigrants cross the border into the U.S. and figured that she would be the right person to get the job done for her brother. That was apparently how it happened: five days before the truck left Harlingen for Houston, Norma and her employee struck a deal: Norma would charge her $1,800 to get her brother up to Houston. The first half was to be paid after he crossed the border, and the other half would be due when he arrived in Houston.

On Tuesday, May 13, when her employee arrived at work, Norma informed her that her brother had made it across the border, and asked for the first payment of $900. But the employee didn't have the money. That same night, Norma telephoned her employee's home and demanded the payment, at which point the employee asked a family member to lend her the money. Finally, she drove to the restaurant and paid Norma the $900.

The other person riding in the trailer that night was Fabiola Gonzalez. On April 30, 2003, Fabiola's grandmother, Josefina Gonzalez, had overheard two women talking at a Houston Fiesta Mart about a woman who was helping their relatives to cross the border from Mexico. Josefina approached the women and asked them how she might contact this person, and according to Fabiola one of them gave her Norma Gonzalez's name and telephone number.

On May 7, Josefina called Norma and asked her if she would be able to get her granddaughter into the United States. Norma said yes, but told Josefina that she would have to wait a few days, because she already had a backlog of immigrants still waiting to cross the border.

Two days later, Josefina contacted Norma again, with better luck this time. That very weekend, Norma would be able to help

Josefina's granddaughter cross the border into the U.S. for a fee of $1,900—a discount off her regular price because Josefina was a first-time client. Usually, Norma explained, she charged $2,000 to take someone across the border and up to Houston.

Josefina accepted the proposal, and Norma gave her a series of very precise instructions. First, Fabiola would have to get herself to the town of Reynosa, Tamaulipas, as quickly as possible. Once there, she was to go to the Hotel Capri. As soon as she arrived at the hotel, Fabiola had to call her grandmother Josefina, who would then contact Norma. Norma, by telephone, would send an individual by the name of "Juan" to pick up Fabiola at the Hotel Capri. "Juan," in turn, would hand Fabiola over to another coyote who would get her from Reynosa to Harlingen.

The plan went off without a hitch. Fabiola crossed the border on Monday, May 12; one day later Norma telephoned Josefina and asked her for $950, half of the fee they had agreed upon. Josefina told Norma to meet her at the Neighborhood Food Store gas station in Houston; there she handed Norma the first payment. The balance would be paid once Fabiola had completed the Harlingen-Houston leg of the trip, slated for the following day.

According to Karla's statement, Víctor Rodríguez, his wife Ema Zapata Rodríguez, and their son, Víctor Jesús Rodríguez, had assembled a group of eleven undocumented immigrants for the truck ride. Among them was a Honduran woman, María Elena Castro Reyes, who had brought her three-year-old son with her. The child, however, would be driven up to Houston in another vehicle.

María Elena and her son had traveled from Honduras to San

Fernando, in the Mexican state of Tamaulipas. That was where her deal ended with her coyote, whom she had paid $1,750 to get to Mexico's northern border. Once in San Fernando, María Elena telephoned Víctor Rodríguez and asked him if he could help her and her son get to the U.S.

On Friday, May 9, 2003, Víctor Rodríguez sent two individuals, who identified themselves as Marcos and "El Chino," to accompany María Elena and her son to the San Fernando bus station. From there, the four of them boarded a bus to the city of Matamoros. Once there, they went to a house that Marcos said belonged to him. That was where María Elena met Doña Ema. None of this came as a surprise. Don Víctor had already told her over the phone that his wife, Ema, would pick up the little boy and personally take him over the Matamoros bridge and into the city of Brownsville, Texas.

María Elena said goodbye to her son, Alexis Neptalí Rosales, and placed him in the hands of Doña Ema. Ema had procured a child's seat for Alexis, both to reassure his mother and to avert the possibility of getting stopped by the police once they were inside the U.S. After tucking him into her green SUV, a 2000 Ford Explorer, they departed. Ema crossed the bridge, and because she is a U.S. citizen, encountered no problems as she passed the immigration service's inspection booths at the border. Ema, born in 1946, could easily pass as the child's grandmother or other relative. What the immigration agents did not know, however, was that Alexis had been born in San Pedro Sula, Honduras, on March 19, 2000, and that he had no blood relation to the woman driving the Explorer.

María Elena had to spend the rest of that Friday and all of

Saturday without her son. According to the plan, she would cross the border into the U.S. that Saturday night and arrive on Sunday. It was hard, very hard for her to be without Alexis. But she agreed: he was too small to cross the border through the Rio Bravo, or Rio Grande, as the Americans called it. The river was far too wide and dangerous for Alexis.

María Elena had no choice but to wait and hope: wait for the moment they would cross and hope that her son was all right. She had never met Doña Ema before, and this concerned her, but she had no other option: if she wanted to go to the United States with her son, this was how they had to do it.

At 9 p.m. on Saturday night, María Elena and two other Hondurans who had traveled with her from their home country—Doris Sulema Argueta and her sister-in-law María Leticia Lara Castro—were led to a spot near the border by Marcos and someone else, someone they didn't know. Once there, they realized that this stranger was in fact the guide who would be taking them across the border and into the U.S. After handing them a couple of inner tubes, he led the group across the river, two by two.

Finally they were on American soil. But the worst part wasn't over yet. The three Honduran women, Marcos, and the guide then had to walk for five hours until someone named "El Canana" finally drove by in a van and picked them up near a neighborhood in Brownsville, just as dawn broke that Sunday morning.

"El Canana" was the nickname used by Víctor Jesús Rodríguez, the son of Don Víctor and Doña Ema, who then took them to a house at 3100 McAllen Road in Brownsville. "El

Canana" deposited them in a room at the back of the house, but at this point María Elena couldn't take it any more. She had to see her son. "El Canana" told her that Alexis was in the house next door, and María Elena went out looking for him. It was late, and she was tired. But finally, mother and son were reunited.

The Rodríguez family had two houses. The one at 3100 McAllen Road was where they hid undocumented immigrants. The one at 3110 was where they lived. In a way, they worked right next door to where they lived.

After resting, the three Honduran women met Don Víctor for the first time. He did not, however, come to the house bearing good news. If they wanted to get from Brownsville to Houston, he told them, they would each have to pay $2,000 more: $1,000 up front, and $1,000 upon arrival in Houston. Don Víctor then asked them for the phone numbers of the relatives who would be making their payments for them. María Elena gave him the number of Alejandro Hernández, a friend of her family who lived in Brownsville and who would be paying for her and her son. When Alejandro received the call, he wasted no time, and immediately transferred the $1,500 to Víctor Rodríguez, to get María Elena and her little boy up to Houston.

On Monday, May 12, 2003, at four in the afternoon, Doña Ema appeared at the house and told everyone to get ready. "El Canana" took the two Honduran women, Doris Zulema and María Leticia, to another house where even more immigrants were waiting. Doris Zulema estimated that there were at least fifty people there—so many, in fact, that the man who seemed to be in charge of the house called Don Víctor on his cell phone to complain that the crowd had gotten too big. Doris Zulema and

María Leticia did not see María Elena until the day they left on the truck.

María Elena and Alexis stayed in the Rodríguez home until Tuesday night. The plan was to drop off María Elena at the truck, and Víctor and Ema would take Alexis to Houston in a green minivan. That was exactly what they did.

Shortly before 10 p.m. on the night of May 13th, Don Víctor and Doña Ema brought María Elena to the outskirts of Harlingen to wait for the truck. Alexis remained behind in the minivan. This was the second time in five days that María Elena would have to say goodbye to her son, Alexis.

According to Karla Chavez, a total of sixty-one people, including María Elena, were sent to the truck: eleven through Víctor Rodríguez, two through Norma Gonzalez, fourteen through Gabriel Chavez, six through Rafa "La Canica," three through "El Caballo," five through Ricardo Uresti, and twenty through "Tavo" Torres. And then there were the people sent by Arturo Viscaíno, Salvador "Chavo" Ortega, and Karla herself. At least four immigrants, one Mexican woman and three Salvadorans, later told federal agents that Karla had helped them get across the border and that, after getting their family members to pay the first installments of their agreed-upon fees, she had promised to send them to Houston.

On Saturday, May 10, one day before Mother's Day in the U.S., Karla and Abel Flores spoke on the phone. Things were moving along well: Abel informed her that they had already gathered sixty immigrants for the trip up to Houston in a few days. All they had to do now was wait for the "trailero," as he called it, to get to Texas. But they hadn't crunched their numbers right: three days later, at least seventy-three people would board that tractor-trailer.

No matter how you looked at it, this was a splendid business for Abel. According to investigative reports, each of the sixty immigrants they were planning to transport would pay him $450 for the ride up to Houston. In other words, he would be making $27,000 in a single day. From that amount, of course, Abel would have to subtract the $6,000 he would be paying Tyrone Williams, plus Williams's expenses. What Abel didn't want to tell the driver was that he actually needed him to drive the immigrants all the way up to Houston. Once Williams was out on the road, though, close to Robstown, Abel would call him on his cell phone and offer him more money so that he would continue up to Houston.

Karla, on the other hand, would be taking in $50 from each of the estimated sixty people who would be traveling, for a total of $3,000. Out of that money, however, she had to pay for the immigrants' food, the time spent in the various safe houses, and all the phone calls she had made to family members demanding payment. Karla calculated that after expenses, she came out with about $500.

But that wasn't the only money Karla was making. She had also received payments from the people she had personally brought over from Mexico and El Salvador; the amounts varied depending on the immigrant's country of origin. The Salvadorans, for example, paid the most: $5,500 for the entire trip to Houston. The Mexicans paid far less, some $2,000 from point to point. Portions of those fees were doled out to pay the various and sundry people who worked for the operation along the way: those who had transferred the immigrants to Harlingen, for example, and then those who delivered them from Harlingen to Houston. Like every business, this one had its ups and downs. But in general, this job was looking all right. Karla would earn money for each and every immigrant that boarded the truck, plus the earnings she had pulled in on her own from the people she had brought over the border.

Karla was not responsible for picking up all the immigrants at the various safe houses in Harlingen and taking them to the spot where the truck was supposed to appear. That was the job of the individual contractors who "owned" their respective immigrants. They had all agreed to meet in the outskirts of Harlingen at 10 p.m. on Tuesday, May 13. According to Karla's statements, Abel Flores would drive his Lincoln Navigator over to the Horizon Motel in Harlingen to pick up Tyrone Williams, who would be waiting for him there. Then, once inside his 18-wheeler, Tyrone would follow Abel's black van to the appointed spot where the immigrants would be waiting. Tyrone was not supposed to see how many immigrants climbed into the truck; there would be many more than he had been told.

Abel would close the two trailer doors before telling Tyrone how to get onto the road to Robstown. The trailer doors would shut, leaving virtually everyone involved in the dark.

4

INSIDE THE TRAILER

The First Hour

Enrique was the last to board the truck. He was still thinking about the old Mexican movie about the immigrants who died inside the water truck. As the doors were shut from the outside, he pressed his body against one of them. Then, he felt the truck lurch into movement.

"Five minutes later I started to sweat, a lot," Enrique remembers. "Everyone started to sweat, it was hot, everyone was sweating and sweating."

After leaving Harlingen, the truck turned north on U.S. Highway 77. The truck would soon have to pass through the town of Sarita, where the immigration service had an inspection booth, and then Williams would head for Kingsville and on to Robstown. When he set out on the road that night, driver Tyrone Williams had no idea that later on he would be asked to continue the journey all the way to the city of Houston, via Victoria, Texas.

The immigrants did not fit comfortably in the truck trailer.

Some people remained on their feet, leaning against the walls, while others squatted, and others sat with their legs pressed against their chests. The metal floors in this kind of trailer are not smooth; they are lined with very thin metal bars that run down the entire length of the trailer, from front to back, to facilitate the manipulation of heavy cargo loads. These bars and indentations make the floor an extremely painful place to sit. Truck trailers are not exactly made for transporting cramped human cargo.

Everyone in the trailer was sweating furiously. It is impossible to calculate the exact temperature, but after a few minutes it might very well have grown as hot as 110 degrees Fahrenheit. As hot as a sauna. Right from the start the immigrants began to experience the first symptoms of extreme heat exposure: dizziness, nausea, an increasingly rapid heartbeat, and an incipient feeling of disorientation. During those first few minutes inside the container nobody lost consciousness, and though they couldn't see it, their skin was growing redder and hotter with every passing second.

The average human body temperature is 98.6 degrees Fahrenheit, 37 degrees Centigrade. Given the increasingly hot temperature inside the trailer, nobody's body was acting normally anymore. When the ambient temperature rises, the human body begins to perspire naturally. When the perspiration evaporates, the skin is refreshed. But conditions were far from normal inside the truck trailer. Inside, the temperature rose at least 40 degrees Fahrenheit higher than the trailer's outside temperature. Subjected to such intense heat, their bodies quickly lost their ability to react normally and regulate their own temperatures. Naturally, the older people and the child were in the greatest danger, because their bodies' self-regulation mechanisms would be the first to fail.

In addition to all this, many of the immigrants were wearing T-shirts, long-sleeved shirts, and jackets. Very logically they had feared that it might get very cold inside the container if the driver put the air conditioning on. Of course, in reality, they were wearing everything they owned—nobody was willing to leave behind the few items of clothing they possessed.

The first half hour of the trip transpired without much conversation, practically in silence. But it wasn't long before people began to grow frantic, according to Israel, one of the passengers.

"People were saying things like 'oh, it's hot, oh I'm thirsty.'" It was so dark that the travelers couldn't even see their own hands, and in the middle of the this dark confusion someone came up with a suggestion:

"Let's take off our shirts and fan ourselves." At that, several men (though none of the women) removed their shirts and began to wave them around in an effort to generate even the tiniest bit of circulation through the heavy, humid air.

This, however, was not enough. The real problem was that there was no way for fresh air to enter the trailer. The situation seemed to call for more drastic measures.

"Everyone in the back of the truck, break the *calaveras* so that we can get some air!" shouted one of the passengers from Pozos, Guanajuato. *Calavera* is the word Mexicans use for tail light. "Break them!"

Enrique and Alberto were at the back of the truck, clinging to the doors. Many of the immigrants had jumped up to their feet, but Alberto had decided to remain seated.

"I was the only one sitting," he remembers. "It was easier to breathe down there, but it was still horribly hot." Alberto, sitting down, saw a shadow cross his path. It was Enrique, who was trying to break part of the door. Alberto went over to him.

Painstakingly, with their fingers and nails, they tried prying off the metal layer that covered the door on its right-hand side, to no avail. Enrique, then, explored the surface of the door with his hands, and discovered that the far end of the door was reinforced with some kind of rubber. He pulled at it, yanking it harder and harder until it finally broke. But then, on the inside of the door, he found another layer of insulation: foam rubber, which he ripped off with his fingers. There, for the first time, they could make out a tiny sliver of light. Frantically, they searched the inside of the container for something to break one of the tail lights. Enrique called out in search of something with which to break the lights. Nobody responded.

"I was pretty desperate by then," Alberto said. "I broke the two lights with my hands, and they were covered in blood, but I didn't even notice until the next day, when I looked down and saw my scars. Enrique hit the lights, too." The two men, with their own bare hands, finally broke the lights. Their hands and fingernails were bleeding, but those two little holes that opened out onto the outside world would save their lives in the end.

"Let me breathe from the top hole. You take the bottom hole," Alberto said to Enrique. The two men placed their noses in the holes and were the first to breathe through them.

The hole, unfortunately, was not large enough to ventilate the entire container. The truck was cruising at barely fifty miles per hour. The driver apparently didn't want to risk getting stopped for speeding by the police, and so he drove well under the legal limit. The hole they had carved out at the back of that container was not big enough to significantly improve the situation for most of the passengers. In order to breathe a bit of fresh air—combined with the exhaust fumes, of course—Alberto had to press his face against the door and stick his nose as far as he could into the hole. The hole had about the same diameter as Alberto's forearm.

"We are going to die in here, I know it." Alberto thought. The holes were not letting in as much air as they had hoped.

"Only more exhaust and more heat came in through the holes. Even so, when the truck hit a curve you could feel a little bit of cold air. I was just happy to feel air on my hand. But everyone kept telling me, 'take your hand out, let more air in.' "

When the other passengers realized the existence of the two holes, they began to rush the door.

"Come on, let us breathe," they said to Alberto and Enrique. Alberto told them they would have to take turns.

"Now you. Now you." There were some people who didn't want to let go once they got to the holes. But it didn't take them long to realize that the air that seeped in wasn't so terrific. About five people, Alberto and Enrique included, stayed where they were, by the two holes they had forced in the door.

The Second Hour

It was just over an hour since the truck had departed Harlingen, and the situation was already dire, especially for the 5-year-old boy inside the trailer, Marco Antonio Villaseñor. His father carried him from the center of the trailer to the back.

"We had already told the father to put the boy in front of the

holes so that he could breathe," Alberto said. Despite the darkness, most of the immigrants could tell that the little boy was in
very bad shape and managed to help the father get over to the
back door so that Marco Antonio could breathe a little easier.

When the humidity level is as high as it was on that May
night—according to meteorological reports, the relative humidity was 93 percent—perspiration does not evaporate so rapidly.
As such, the body will not cool off and its temperature will not
go down. That night, the immigrants felt as if their entire bodies
had been coated with some kind of slick, slightly greasy sub-
stance. Not only were their clothes completely soaked through,
but they could even feel water swishing around inside their
shoes.

Despite the heat, the passengers' mouths were extremely
dried out, and the truck's movements gave many of them
vertigo—a sensation that some of them had never experienced
before. They began to grow disoriented, and after a certain point
it wasn't so easy to know which way was up and which way
was down.

Then, of course, there was also the psychological element.
By their second hour inside the trailer, many of the immigrants
were hyperventilating. The darkness, the fear, and the strange
noises made by their fellow passengers only added to their
anxiety, and their breathing grew faster and more labored.

Their short breaths made them feel as if their thoraxes were
pressing against their lungs; this is the body's natural reaction
when it perceives a lack of oxygen. Hyperventilation and fear,
then, became a vicious and fatal cycle inside the trailer. Some
people began to lose consciousness, and some felt the need to
vomit.

They had no way of measuring these things, but their bodies had grown hotter than 100 degrees Fahrenheit and were dehydrating rapidly. In general, human body temperature can go up 5 or 6 degrees in less than fifteen minutes. But in addition, they had no water to drink, and after so much perspiring, their skin suddenly had no more liquid to draw from, and yet the humid air kept them coated in a layer of sweat.

Everything seems to indicate that after an hour and a half on the road, as the truck approached the immigration service's checkpoint booth in Sarita, Texas, nobody had died yet. But it would only be a matter of minutes before some would begin to perish, one by one.

Inside the trailer, the immigrants began discussing what to do once the truck pulled in at the inspection booth. They could make a lot of noise and bang against the walls so that people would hear them. That would certainly get someone to open the door—but once the door opened, their journey would end. Not only would they almost certainly be deported but they would also lose all the money they had already paid the coyotes. Somewhere in the darkness a voice piped up, saying that in half an hour they would all be let out of the truck. At that, the debate ended. Their voices once again falling silent.

There was no way they could ever guess that the trailer doors wouldn't open for yet another two and a half hours.

Doris Sulema Argueta, the Honduran woman, was traveling with her sister-in-law and a friend, also from Honduras. One of the first things that caught Doris's eye was a little boy, about 6 years old, she guessed, who cried and cried. His father had removed almost all his clothing so that the heat would be a little more bearable. According to Doris, the child did not survive much longer.

Doris was also standing near a man who was vomiting. Then, suddenly, he stopped. But not because he was feeling better. Doris surmised that, like the little boy, the man vomiting next to her had probably died.

It seems that only one person inside the trailer was carrying a cell phone: Matías Rafael Medina, a 25-year-old Honduran who did not speak English.

"I have a telephone with me," Matías called out. Alberto heard him. "I'm going to try to make a call." At first, the phone had no dial tone but after a little while, Matías was able to place a call, which he did: 911.

After several attempts, an operator answered.

"They answered me in English," Matías later told reporters. "And since I don't speak the language, we didn't get any help." That phone call, which was one of their few precious chances to obtain assistance, failed because the call was not transferred to a Spanish-speaking operator.

In Texas, one out of every three residents is of Hispanic origin and, as such, it is very likely that the people who fall into this category speak at least some Spanish. But that was not the case with the operator who answered Matías's phone call. Apparently, some kind of attempt was made to transfer the call to someone who spoke Spanish, but the connection was lost and they were unable to trace Matías's cell phone number.

Meanwhile, according to investigative reports, Karla, Abel, and the driver were in constant contact with one another, completely unaware of what was happening inside the truck container. Karla Chavez had two cell phones. One was for her, and the other was for Abel. The numbers were very similar. Karla's phone number was 956-357-0242, just one digit away from the number on Abel's phone, which was 956-357-9242. As such, it was almost impossible for them *not* to be in contact. In fact, ever since the truck had left Harlingen, Abel and Karla talked almost nonstop, with Abel giving Karla periodic status reports. Abel knew what was going on in the truck because he was also in constant touch with the driver, Tyrone Williams.

For Tyrone, the deal was pretty simple. All he had to do was drive the immigrants to Robstown, Texas, which was halfway between Harlingen and Houston, taking the road that hugged the Gulf of Mexico coastline. For this, he would earn $3,500. But once he was out on the road, Williams received another phone call offering him $2,500 more if he agreed to drive all the way to Houston. It was tempting. Williams would earn $6,000 in less than eight hours. Without thinking twice he said yes. Six thousand dollars was almost double what he earned for a typical coast-to-coast delivery job. Plus, they had promised him that there wouldn't be any trouble with the immigration service inspections. And they made good on their promise: as he drove past the checkpoint, his truck was not stopped for inspection. All would have been different if Williams had been.

The truck didn't come to a full stop anywhere in the immigration checkpoint area. Williams simply slowed his vehicle down to a crawl and then continued on his way. The photographs of the truck as it passed through the checkpoint—which were later included as part of the evidence in the case against the fourteen people accused of trafficking undocumented immigrants, among other charges—reveal at least one hole on the back side of the truck. Why wasn't the truck inspected? Why was it allowed through without anyone inspecting its cargo? Didn't the hole in the door raise any eyebrows? Were there enough employees on duty that night to properly inspect all the trucks and cars on the road at that hour? It seems rather obvious to point out that if the driver and the coyotes picked that particular route to drive up to Houston, it was because they were confident that the truck and its cargo would not be inspected.

Alberto passed the time touching a scapular that he had bought at the Basilica of Guadalupe in Mexico City before coming to the United States.

"I put myself in the hands of the Virgin, so that she would give me the strength I needed to live through the hell I was in," Alberto remembers. His long, black, shoulder-length hair only intensified the heat he felt.

Inside the container, the situation was getting worse by the minute. Alberto just couldn't keep dry.

"I would wipe the sweat off with my shirt and then wring it out," he remembers. "And that very same moment, just as I finished wringing the shirt, I would have to wipe myself off and wring it again, and this over and over again."

Alberto ended up drinking his own sweat. Running his cupped hand across his chest, he gathered up a little bit of liquid in the center of his palm and drank it. That was when it first occurred to him that he was going to die.

"I began to regret all the bad things I had done, so that God would forgive me," he remembers. "You could hear lots, lots of people talking, lots of people praying, almost everyone was asking God to forgive them. I did the same thing, because I knew we were going to die. I knew I was going to die because it was hell in there. It was awful. I had had nightmares before, but they were nothing compared to this." But as he confessed out loud, Alberto kept on fighting.

Through the hole in the back of the truck, Alberto and Enrique began to signal to the other cars on the road. "We stuck our hands out, we threw clothing, shirts, baseball caps, everything we had. We shouted our lungs out through that hole." During those three hours, Alberto and Enrique thought their signaling had gone unnoticed. But they were wrong.

Matías was not the only person who had telephoned the police that night. In the city of Kingsville, just north of the inspection booths in Sarita, a man saw a truck with New York license plates go down Corral Street just before midnight. Something was very wrong. He saw a hand waving something that looked like a scarf or a bandanna through a hole in the back of the truck. That was enough to make him call the Kingsville police department. The operator took his call, but the call was not handled as an emergency.

The Third Hour

Alberto believes that a car or a minivan was escorting the truck for almost the entire ride, because he kept on seeing the same large lights from a vehicle that seemed to be following them.

"A normal car would have reacted [to the signaling]," he mused. "But the people in that car did not."

Thanks to the sliver of light that entered through the holes in the back of the trailer, Enrique could see a little bit of what was going on inside the trailer. What he saw was more terrifying than the just prior absolute darkness.

"I saw everyone bathed in sweat," he remembers. "The trailer was completely white from all the steam, it was like we were in a steam bath, or some place with hot water coming out. It was like [when there is] fog, and you can't see very far into the distance."

Enrique's eyes began to burn, and he felt certain he was about to die.

"At one point, my vision got completely blurred. And then I saw something that looked like a person with his head all covered up. You couldn't see his face or his body. Nothing. It was like a body covered in a white sheet. I thought I was going crazy. I thought I was dead, that all these things were part of the afterlife. But at the same time, I could still hear the other people in the truck talking. That was for a minute, maybe two. Then, just as suddenly as before, I felt as if someone was lifting me up off the floor and then suddenly I came back to my senses, and I realized I wasn't dead."

Enrique hadn't died. But some of his fellow passengers were indeed perishing from the lack of oxygen.

"There were people lying on the floor, not moving at all." Enrique could feel them with his feet as he tried to move from one end of the container to the other. "I couldn't imagine they were dead; I assumed they had fainted."

Enrique was not the only person in the truck who was hallucinating from the lack of oxygen and from breathing in too much carbon dioxide.

"I stood up, and then sometimes I sat down, but I couldn't stretch out at all because there were too many people in there," said José, who hadn't brought his diabetes medication with him. That didn't worry him, not for the moment, at least. He was, however, concerned that he might have an asthma attack. That was a luxury he couldn't afford at that moment, because it would very probably cost him his life. José knew he had to save his energy, so he stopped talking and focused on not spiraling into desperation.

José was sitting next to a 17-year-old from El Salvador. With a baseball cap he fanned the two of them. A puddle of sweat had formed beneath José, who said that the Salvadoran teenager "stuck close to me and said, 'help us.' " Together, they counted "one, two, three" and began taking deep breaths in unison. They repeated this exercise several times.

"I can't, I can't, I'm not going to make it," the Salvadoran boy said to José. But José kept on encouraging him:

"Come on, man, we're almost there, we're almost there."

For José, the real problem was not the heat coming from his Salvadoran neighbor. Lots of people around him were on their feet, and he was getting soaked from their perspiration, which was dripping down on him, so he decided to stand.

"Oh God, I'm going to have to get up," he said to himself. "Their sweat is dripping all over me." The droplets that fell on José's head, however, did serve a purpose, for they kept him alert and awake.

José had worked outdoors in the country and had performed many jobs that required tremendous physical effort, but he didn't have much flexibility or muscle tone. Illness and inadequate medical care had taken their toll on this man, who was still under 50 when he stepped into the trailer. With great difficulty, José pulled himself up and leaned against the trailer wall, successfully moving himself away from the falling shower of perspiration above him.

"I wanted to vomit," José remembers. "I felt a kind of desperation." Then he began to walk, taking very little steps. But as he got up, he again realized the several people lying on the floor were no longer moving.

"Get up, man, why are you going to sleep?" he shouted at them with what little strength he had left. "Get up, man! Look up! Don't go falling asleep!" But nobody seemed to pay him any mind. José's urge to vomit was now under control, but he still felt distraught. He went back to his previous spot and sat down on the floor.

José says he remembers hearing one man confess to him, "I drank my urine." But nobody else in the truck seems to remember having heard anything of the sort. Suddenly, José heard a loud belly laugh that frightened him. At first he thought it was the sound of a little boy laughing, but it didn't sound normal. "It sounded ugly," he said. Later he admitted that it was a sound he'll never forget, the most prophetic sound of his life.

By then, breathing had become a serious challenge for everyone. Their nasal passages would flutter open in a futile attempt to take in more air. This nostril flapping was then followed by grunts and very hoarse noises as they tried to exhale. That was the sound of their bodies struggling to keep their lungs open. Under normal circumstances, a person breathes in and out some 25,000 times a day, but by this time the immigrants' breathing rate had risen considerably higher than that. Many of them felt themselves starting to lose consciousness.

Hyperthermia, or high body temperature, made them feel as if they had just depleted all their energies in an exhausting long-distance run. By now, most of them were terribly weak and their pulses, barely perceptible. Still, some of their bodies reacted violently. Disoriented and wobbly, several passengers pushed one another as they tried to feel their way toward the back of the truck in the hopes of getting a tiny bit of fresh air. Some people had become delirious, while others fell into comas.

Ironically, in the middle of that darkness many of the immigrants felt as if they were suffering from something like sunstroke, as if they had played two or three games of soccer beneath the piercing Mexican sun. But neither sun nor soccer balls were to be found inside the truck. Just the agonizing heat. And not even a little bottle of water.

One of the greatest mysteries of this tragedy is why none of the women died. There is no doubt that women have a higher pain threshold than men. But male and female bodies are the same in that they are both made up of 70 percent liquid. What was it that helped the women survive, while so many of the men perished? Was it their psychological response to danger? Were they physically stronger and more resistant than the men? Were they better at storing their energies and not spending their energy by uselessly complaining?

José had another, less rational explanation: the women survived because they immediately began praying. Next to him, some Salvadoran women began praying:

"Glory to God!" they cried. "Halleluja, we are going to be saved, don't worry. Halleluja!"

José, right then and there, was not so convinced that they were all going to be saved. On the contrary: he suddenly heard a man call out to the devil, which scared him even more:

"Satan," José heard someone call out. "I made a pact with you. But my God, if you let me [live], I promise I will serve you and only you." Prompted by these diabolical dialogues, someone else sitting near José began to pray.

"The Lord is my shepherd, I shall not want." José, unable to resist the call any longer, joined in with the stranger sitting next to him, and repeated the words,

"The Lord is my shepherd, I shall not want."

José is convinced that it was God who "let me live."

The policeman who had stolen José's money in Reynosa had not taken the image of the apostle St. James that he carried in his wallet.

"Protect us! Protect us!" José prayed, almost shouting, as he patted his trousers in search of his wallet. "Father, forgive us. Forgive all of us in here because we are not worthy to be with you [in Heaven]. Oh, Lord! We don't deserve to be with you because we are all sinners! Forgive us!"

Israel, José's nephew, was not just scared but also quite uncomfortable because there was a woman stretched out across his feet.

"I'm sorry to bother you, but could you move over a little so that I can rest my feet? I'm real tired," he said to the young woman. "But then she said to me, 'I don't feel too good myself.' "

Israel had taken the advice of his uncle and removed his shirt, but that didn't stop him from sweating. "It was so heavy from sweat that it slipped out of my hands. And the heat from everyone else, that suffocating feeling, it was unbearable." His socks were soaking wet and his boots felt "as if they were under water."

Hoping to stretch his legs, Israel tried standing up, but it wasn't so easy to keep his balance, so he sat back down. That was when he realized it was easier to breathe down by the floor, where he would tumble around and bang his hands against the trailer doors. Just like all the other immigrants, he was hoping the driver would hear their noises and open the door. But all he got for his efforts was a banged up, bloody hand.

Before leaving Pozos, in Mexico, Israel had asked his girl-friend for a blessing. And for the trip itself he had brought along the images of two holy men, which he kept tucked away in his billfold: one was of Rey Justo Juez, venerated in Pozos, and the other was an image of Jesus Christ with an inscription that read: "God of the Ocean."

As time went by, Israel began to notice that several of his fellow passengers had stopped moving, and he finally began to suspect that many of them must have died. He felt himself go in and out of consciousness.

"It was like I was slowly dying in there, that was how bad I felt." Then he went over to one of his uncles. Israel says he felt "sad" and "alone" when he realized that there were dead people all around him.

"Oh God, it can't be true. We can't die inside here," Israel thought. Sitting in the middle of the truck, he was able to see the back door thanks to the light filtering in through the tiny holes, but he was afraid to look too closely, because it seemed that near the door was a pile of dead bodies, people who had tried to breathe through the hole.

The effort had been too much for some of them.

"I saw death," Israel declares, as the tears threaten to spill from his eyes. "I saw it, very clearly, right there in front of me."

"In the darkness I saw a shadow pass by me, all of a sudden," he remembers. "And I said: the devil has taken over this truck . . . so many people cursing. So many people shouting all kinds of things to the devil, to Satan. I felt awful, really awful, and all I could say [was]: 'God, if I die, take me quietly.' " Israel tried to get up and move toward the back of the truck, but he was suddenly struck by the feeling that he too would die, just like the others had before him, if he moved from where he was sitting. His energy depleted, he lost consciousness. After some point, he doesn't remember a thing.

The worst moment of all was when Israel heard one of his uncles say that he was about to die.

" 'My nephew is going to die here, too,' " Israel heard his uncle say. "Maybe he thought I was already dead," Israel muses, with a pang of guilt, for he was unable to save his uncle's life. "He tried to save me, and I did everything I could for him, too."

Alberto, a few steps away from Israel, hadn't given up yet.

"I think my will to survive was what saved me," Alberto reflects. "More than shouting, or crying, I said, 'Lord, I know we are going to die, but I want to die fighting. We really are going to die.' "

"He's dead now!" Alberto suddenly heard someone say. "No! Hit him, in the chest, so he'll react." He couldn't see who they were talking about.

"I started to fall asleep now and then," Alberto said; his strength was dwindling. "That was when I really got scared."

"Lord, I think we are going to die in here," Alberto prayed. "Forgive me if I have failed here." Alberto felt the end was coming. But he wasn't the only one praying. A group of women in the truck were busy praying as well.

"I remember it, as soon as things began to get desperate, they all began to pray," Alberto muses. "I had never heard those songs before. They sang. They prayed. Instead of wasting their energy they just prayed."

"You're all hot," Enrique said to Alberto. "Get off me." Both of them recoiled whenever their hot skin came into contact. But Enrique's remark succeeded rousing Alberto from the lethargy that had overcome him.

"I was having a dream, and I felt so nice, so good." But Alberto jumped up again, this time to stick his shirt out of the hole in the back of the truck and make noises against the truck's exterior. Everyone was either nauseated, disoriented and, like Alberto, on the verge of passing out or already dead. But the survival instinct is the strongest instinct of all, and it helped keep at least some of them from falling into the sleep that would inevitably lead to a loss of consciousness and then death.

Some of the passengers had begun to feel cramps in their extremities and a strange, uncomfortable sensation around their mouths. So much sweating had depleted their bodies of salt and this, in part, was what brought on the cramping. Due to the excessive, constant heat, a rash of little blisters had begun to break out on their necks and hands. But when they tried to scratch at the rash, their fingernails turned black from the combination of sweat and damp skin cells. Their breaths per minute continued to rise as their bodies continued reacting to the lack of oxygen. Some say they had an easier time breathing down by the floor, while others say they preferred to remain standing. In the darkness, there was no way they could tell, but their skin had already turned an extremely pale, grayish-bluish hue.

The passengers' feelings of despair only exacerbated their physical conditions: as a general rule, when breathing quickens, the heart must work much harder than normal, at a level of effort that is impossible to sustain for very long. If someone didn't get them out of there very soon, they were all going to die.

The intense, collective body heat began to claim more victims. Several passengers' bodies' lost their ability to regulate their temperatures, and this gave way to tremors, convulsions, swelling of the lungs and, finally, sudden death brought on by arrhythmias or heart failure.

The word *asphyxia* comes from the Greek (*a*, not; *sphyzein*, to beat). Literally, it means "to stop" or, in other words, to be left without a pulse. This is precisely what happened to the majority of the immigrants who were, by then, dead.

During the normal breathing process, air travels through the nose or mouth toward the trachea and then on to the bronchial tubes and the alveoli. For many of the people inside the trailer, this process had come to a complete halt. The lungs contain some 700 million alveoli, tiny sacs of tissue where oxygen and carbon dioxide are exchanged between the lungs and the blood. Through the alveoli, the blood carries the oxygen to the heart and bodily tissues, while the lungs exhale the carbon dioxide. The body is a machine that functions perfectly, so long as it is not exposed to external factors, such as intense heat. A few degrees' difference in temperature can upset this marvelous balance and cause death.

Little by little, millions of alveoli in the lungs of several passengers ran out of the oxygen needed to exchange with blood, causing a grotesque death scene which, for better or for worse, nobody would be able to see through the blackness inside the trailer. Life, in the end, hinges on those alveoli and their ability to perform their simple function: injecting small bubbles of oxygen into the blood. Life begins and ends in the alveoli. If those millions of communicating vessels stop functioning, life stops.

The dead were stretched out across the floor, strewn among those who had fainted and those who had fallen asleep, but there was no real way to tell who had died and who had just passed out. The bodies of the dead continued to give off a great deal of heat—106 degrees Fahrenheit, maybe more. For this reason, many of their fellow passengers did not realize that some around them had died. As they neared death, some passengers suffered from muscle spasms, belches, and the passing of gas, and on occasion these functions actually continued for a few moments after death. This also led their fellow passengers to believe that they were still alive. They were not.

The difference between the living and the dead was a breath, a prayer, a monumental effort to keep the eyes open. Nobody was crying anymore; their bodies didn't have enough liquid to form tears. If the humidity inside the truck hadn't risen so dramatically, their dry skin would have cracked and bled due to the lack of bodily fluids.

Panic and anxiety attacks further worsened things for some of the immigrants, who might have been able to survive had they been able to hang on to a bit more energy. Hyper-anguish, so to speak, was exacerbating the hyperthermia and hyperventilation they were already experiencing. Unless they could break that cycle—and fast—there would be no escape for them.

As all this was happening, the people who had still managed to hang on to a bit of strength remained glued to the walls of the container.

"¡*Párate ya!*" they screamed at the driver in Spanish. "Stop the truck! Stop the truck!"

The survivors interviewed for this book say their banging efforts were loud and strenuous and they find it difficult to believe that the truck driver did not hear them banging away, as they prayed and cried out to God, and some, in their desperation, to the devil. Some passengers even tried to turn the truck trailer upside-down, by making everyone lean toward one side of the truck at the same time. But the experiment failed; the trailer didn't budge, and the truck just continued on its way.

Walking around inside the trailer was not easy, either. Several passengers say that the truck would sometimes stop short and then accelerate. These sudden, brusque movements may have been produced by the conditions out on the road, but they also may have been the driver's attempt to get them to stop leaning against the trailer walls. There is no way to know for sure.

The Final Hour

Just after about three hours on the road, the passengers finally began to feel the air conditioning system kick in. Some of the survivors say that they felt "a bit of air" waft across their over-heated bodies. Doris, the Honduran immigrant, felt it. But the remedy had arrived too late. A person suffering from hyper-thermia and a body temperature of more than 100 degrees Fahrenheit will not be able to recover with air conditioning alone. A high body temperature, for one thing, will not go down on its own for several hours. And time was precisely what these people lacked: it was running out with every breath they took.

Whistles seemed to echo through the air during this fourth hour of the journey. Who could possibly be whistling under these circumstances? Yet these were not normal, everyday whis-tles, like those of a little boy. They were the sharp, almost chirp-ing sounds made by the passengers' narrow, swollen windpipes as they breathed the air inside the truck.

A mixture of saliva and blood trickled from the mouths of some of the dead. Although their alveoli were now unable to ex-change much oxygen with the blood, they probably would not have torn apart as alveoli often do under more extreme circum-stances such as fires, when people inhale boiling air and smoke. The blood vessels in the stomach and the breathing passages, however, are not quite as resistant as the alveoli. Without a doubt, they were the cause of the bleeding. Some passengers said they experienced intense stomach pains, similar to those of an ulcer, which were followed by expulsions of saliva loaded with blood clots. Those who survived were now facing the possibility of brain damage or acute kidney failure.

Then, something changed.

After four hours on the road, the truck suddenly came to a halt. For the first time, it came to a full and complete stop. Driving along U.S. Highway 77, driver Tyrone Williams had cruised past Robstown, Refugio, and McFaddin, but as he approached Victoria, Texas, he apparently noticed that one of his tail lights was dangling. Up ahead he saw a gas station on the side of the road and pulled over. It was the Hilltop Exxon Truck Stop.

As soon as he stepped down from the truck to examine the tail lights, Williams heard people shouting and banging from inside the truck container.

Enrique could now see the driver through the hole he had made by punching through the trailer's tail lights.

"Who is it?" Alberto asked Enrique. "A Mexican?"

"No," Enrique answered. "A black man."

"You have to tell him to open the door."

"Open the door," Alberto cried. "Open the door!" Next, Enrique tried speaking to driver Tyrone Williams in very rudimentary but straightforward English.

"I looked at the truck driver, and I got the feeling he was unhooking the truck container from the driver's cab," Enrique recalls. Inside the trailer, the passengers who were still alive continued banging against the walls so that he would open the doors. Enrique heard the driver say that he was going to leave them there and gave them no indication that he was going to open the door.

Enrique was distraught.

"Excuse me, man," he cried out in English. "No more water, the baby is die!"

When the driver heard that they didn't want more water and that a baby had died, he grew alarmed.

"One guy died?" Williams asked, not fully convinced that this could possibly be true.

"Yes man, guy die," Enrique insisted, again in English. He told Williams that at least one person had died and that if he didn't open the doors soon, they would all die in five or ten minutes and that he, Tyrone, would be in trouble with the law. "Open door, please, you no open the door maybe is die everybody in ten minutes, five minutes. You gotta problem with the police."

"What?" the driver replied. He repeated the same question as before. "Guys are dying?"

"Yes, man," Enrique said.

Four hours after departing from Harlingen, Texas, the driver pulled a lever and opened the truck container's two doors.

"The two doors opened," Alberto remembers. "People fell down. I felt a little weak, and said to myself, 'I think I'm going to stay here a little bit.' But then I got scared that they were going to close the doors again. There were a lot of people on the floor."

Tyrone Williams suddenly found himself staring at several people in fetal position; that was when he first realized something was very, very wrong. He kept hearing a woman cry out, over and over again.

"*El niño, el niño . . .*" Williams does not speak Spanish but he rapidly made the connection between the woman's cries and what Enrique had told him. Indeed, it seemed that a minor may have died inside the truck.

All of this sent Williams into a state of panic.

At 1:55 a.m., Williams decided to enter the gas station convenience store; a security camera caught him on tape just as he approached the cash register to pay for twenty bottles of water. But it wasn't enough.

Williams distributed the bottles of water among the immigrants. As soon as they caught sight of them, many of the passengers pounced. Some bottles exploded in the chaos, grabbed too tightly by too many anxious hands. The people who managed to get their hands on a bottle downed their contents in practically a single gulp. Enrique secured a bottle for Alberto and passed it to him.

"I drank a little and then hid it under my clothes, like it was my one and only treasure."

"There was a girl near me, I touched her neck," recalls Alberto. "She was still breathing. I grabbed the bottle I had hidden and poured some water on her lips. Back then I had long hair, and she grabbed it, totally desperate, and refused to let go. I pushed her away. Then someone came by and grabbed the water bottle in my hands away from me. 'The water is for the girl,' I yelled, but he didn't understand. Then I heard the truck make noises, as if it was about to take off. I got down right away, but the girl stayed where she was."

In the middle of all this, Fatima Holloway, an African-American woman wearing a striped, sleeveless T-shirt and capri-style pants, entered the convenience store, where she was captured by the surveillance cameras as she made a purchase. It was 2:03:47 in the morning on Wednesday, May 14, 2003.

Her movements indicated no sense of urgency whatsoever.

She moved back and forth between the purchase counter and the shelves loaded with bottles of water. Almost eight minutes went by before she emerged with two white bags. By then it was 2:11:33.

Following this, a very frightened Williams, according to his own statements to investigators, disengaged the driver's cab from the truck bed and fled the scene with Fatima Holloway. The trailer was left abandoned on the side of the road, as were the people in it.

5

THE LITTLE BOY

A funny thing happened the day he was born, March 10, 1998," said Carolina, the mother of Marco Antonio, the only child traveling in the truck. "When he came out, he didn't cry. They took him away to inoculate him, and all the babies cried except for him. He opened his mouth, yawned, and went back to sleep. The doctor who treated him, the pediatrician, said 'Oh, aren't you a little sleepyhead.' That same day the nurses took pictures to see which was the biggest, chubbiest baby. It was Marco Antonio."

"He was so sweet, so friendly, he could meet anyone and talk to anyone about anything. He was 2 years old when he first saw that other children went to kindergarten, and since he was an only child, he said to me, 'I want a little brother.' And I remember saying to him, 'No darling, we can't afford one, not yet.'" The day Carolina brought him to school for the first time, she was the one who cried.

"He saw me crying," Carolina recalls perfectly, "And he said, 'I don't want you to be lonely—if you want, I'll go home with you.' And I said to him, 'No, sweetheart, you stay here.'"

Marco Antonio liked to look at the *Spider-Man* books and color them with crayons. He had lots of nicknames, too. His uncles called him "Angel," or "Angelito." His mother liked to call him Maclovio.

"He liked guns," she explained. "Maclovio" is the name of a song by the popular Mexican singer Vicente Fernández, about a man named Maclovio who likes to play with guns.

"Do you like that, 'Maclovio'? " she asked the little boy.

"Yes," he replied.

"So I called him Maclovio," Carolina explained, although Marco Antonio himself couldn't quite pronounce it and always said "Covio" instead.

Marco Antonio had flat feet, a condition that required him to use insoles from the age of two. His last pair, fitted before he left for the United States, were made of metal. He didn't suffer from asthma or any other respiratory illness, and his doctors commended Carolina for keeping her son's vaccinations for diseases like pneumonia and hepatitis up-to-date.

Marco Antonio's life, however, was not easy.

"My husband and I had our differences," Carolina confesses. Their differences eventually became so great that the couple separated after living together for seven years; they had never married. She made the final decision to leave his father, José Antonio, on Tuesday, February 25, 2003, when, as she recalls, he struck her in front of the child. The dispute began when José Antonio, a taxi driver, asked Carolina to help him fix the brakes on his taxi. She replied that she couldn't, because she had to prepare dinner.

"That was what got us arguing," Carolina recalls. "He hit me." When the little boy saw how his mother had been struck in her right eye and thrown to the opposite side of the room, he said to his father,

"Don't play like so rough, *cabrón.*"

According to Carolina, José Antonio turned to the child and said, "You stay out of this," and then physically lashed out at him, too.

"That was it for me," said Carolina. "Me, out of love, or out of whatever, I could let it go. But I said to him, 'Don't you lay a finger on my child.' " José Antonio, it seems, then asked Carolina to kiss him. She refused, and shortly afterward José Antonio stormed out of the house. Carolina didn't even waste time packing her clothes; she just grabbed Marco Antonio and took him to her parents' house. She didn't speak to José Antonio for several days after that.

Carolina Acuña was 28 years old at the time. She comes from a large family, with lots of children: Víctor Acuña and María Martínez had eight children and raised them all in Lázaro Cárdenas "La Presa," a neighborhood in the municipality of Tlalnepantla in the state of Mexico. After her dispute with José Antonio, Carolina didn't doubt for a second that her parents' house would be a safe haven for her.

Two weeks after the violent incident, José Antonio called Carolina's parents' house, wanting to speak with his son. After their second phone conversation, Marco Antonio said to his mother,

"I want to see Toño, I miss Tonín." That was what he called his father.

José Antonio kept urging Carolina to come back home with Marco Antonio.

"Quit playing your stupid games," José Antonio said to Carolina. "Get back here, start the wash, and get going on the housework." But Carolina refused.

"No way," she said to herself. Now less than ever did she want to go back with him.

But Carolina could not keep Marco Antonio from seeing his father. Often, when she went to work, she would leave him in the house she had previously shared with José Antonio. On Thursday, March 13, however, José Antonio got into his taxi and drove over to Carolina's parents' house looking for his son.

"I got very nervous. He scared me," Carolina recalls. "As Marco Antonio got into the car, José Antonio looked at me and said, 'I'm really going to give it to you now.'"

José Antonio, once again, tried to convince Carolina to come back to him, to get in the car. When she refused, this is what she says happened next:

"He twisted my hand backwards and then hit me and broke one of my teeth. And I stood there bleeding, my nose, my face." Marco Antonio, apparently, had crouched down in the front seat of the taxi so that he wouldn't have to see his mother being beaten. José Antonio then got into the car and took off, and Carolina undid her ponytail so that her hair would cover her face, to keep people from seeing that her nose was bleeding. After that incident, Carolina would not see her son for almost seven weeks.

Emma Villaseñor, also known by her nickname Mimi, would later tell a *Washington Post* reporter that Carolina had in fact left the house on her own, and that she placed Marco Antonio in the care of Mimi's brother José Antonio.

"I couldn't see him, but I wanted to," Carolina says, referring to Marco Antonio. Every so often, she was able to speak with him on the phone.

"I was always talking to him, saying, 'I love you . . .' all the time," Carolina recalls. But according to Carolina's recollection of things, José Antonio always managed to come up with some kind of excuse to keep her from seeing the boy.

On Tuesday, April 29, 2003, while talking on the phone, Marco Antonio said something that frightened his mother:

"Mommy, I want to see you because I'm going to the United States in a few days," the little boy said before someone cut the connection. Carolina, who on other occasions had heard José Antonio talk about wanting to go and work in the United States, suspected the worst.

"Something is very wrong," she thought. The next day, April 30, the Day of the Child in Mexico, Carolina went to see her son in the house she and José Antonio had once shared. She knocked on the door, and Marco Antonio answered. She knelt down and hugged him; they hadn't seen each other in forty-seven days—since the day that José Antonio had attacked her. Marco Antonio had toys and candy with him. Then, all of a sudden, Carolina grabbed her son took off running with him, an impulse she hadn't planned.

"I picked Marco up and ran, I just ran away," Carolina recalls, as the tears roll down her face. "But one of the neighbors called out 'Caro is kidnapping Marco! Caro is kidnapping Marco!' I wasn't committing a crime—Marco was my son. He was my son, my only son, my little friend! I ran and ran and ran. His feet, in his orthopedic shoes, banged against my legs, giving me bruises. My legs hurt. Then Marco said, 'Look, here comes Toño,' and that made me run even faster. I always wear my hair in a ponytail, and suddenly he grabbed me by the hair and threw me against the sidewalk, cursing at me. I shielded Marco

Antonio with my body to protect him, and then I fell onto the sidewalk and José Antonio hit me and hit me until he finally yanked Marco Antonio away from me. I could feel him beating my face, I could feel the kicks, the curses. That was how he got me to let go of Marco. The other women who lived on the street realized what was going on and they shouted, 'Let go of the child! Let him go, please!' And Marco said to him, 'Daddy, stop it, stop it, that's enough.' After a few more seconds I let go of Marco so that he could go back home.

"[José Antonio] turned to me and said, 'See what you did?' And I said, 'I want to see my child.' 'You could have said so,' he replied. 'I always do but you never let me see him.' 'Well I have things to do,' José Antonio told me. Then he took Marco Antonio, and muttered to me, 'You so-and-so, I'm really going to give it to you now.' That was when I knew: 'He's going to run away with him,' I said to myself.

"After that I went to Ciudad Neza to press charges against him, and right then my eye was in pretty bad shape—I may have ended up with a detached retina from when he hit me. I mean, I looked worse than a prize fighter. I pressed charges in the town hall [of Ciudad Netzahualcoyotl], but by the time they processed and registered the claim, he had already gotten away with it. He had already taken off with Marco."

After that day, Carolina would never see or speak to her son again.

One week after the last beating, May 8 or 9 (she doesn't remember exactly), Carolina summoned up her courage and telephoned one of José Antonio's family members.

"Don't worry," Carolina was told. "The boy came over here

One of the first images of the immigrant tragedy that took place in Victoria, Texas, May 2003. At least seventy-three people, mostly Mexican, were riding in the trailer. *(Courtesy of Univision)*

Karla Chávez, whom prosecutors identified as the ringleader of the trip that caused the death of so many undocumented immigrants, was arrested after she fled the United States, entered Honduras, and then attempted to travel to Guatemala. She was first extradited to Honduras via Houston, where she was met with criminal charges. In the end, the prosecutors did not ask for the death penalty in her case. Karla Chávez pled guilty to one of the fifty-six charges against her in the hopes of receiving a reduced sentence. *(Courtesy of Univision)*

An image of truck driver Tyrone Williams following his arrest. Williams was charged with transporting undocumented immigrants and fleeing the scene of a crime. He was one of the defendants for whom the death penalty was sought. *(Court document courtesy of the State of Texas)*

Reporter Jorge Ramos *(at right)* with four of the survivors. At least fifty-four immigrants survived the tragedy. *(Courtesy of Univision)*

Marco Antonio Villaseñor Acuña, the only minor who died inside the trailer. *(Courtesy of Carolina Acuña, mother of Marco Antonio Villaseñor Acuña)*

Marco Antonio, just after his fifth birthday, with his dog in Mexico City. (*Courtesy of Carolina Acuña*)

The child with his mother, Carolina Acuña, in Mexico City. (*Courtesy of Carolina Acuña*)

Marco Antonio with his father, José Antonio Villaseñor, who also died inside the trailer, and his mother, Carolina Acuña. *(Courtesy of Carolina Acuña)*

Carolina Acuña, in Mexico City, following the death of her son. *(Courtesy of Carolina Acuña)*

Carolina Acuña taking flowers to her son's grave in a Mexico City cemetery. *(Courtesy of Univision)*

Four of the survivors inside a trailer similar to the one in which the tragedy occurred, as part of a special Univision television program. *(Courtesy of Univision)*

From left to right: Surivivors Alberto Aranda, Enrique Ortega, Israel Rivera, and José Reyes sit inside the trailer *(Courtesy of Univision)*

Alberto Aranda and Enrique Ortega in Houston, awaiting the trial of Karla Chávez. All the survivors were given a special visa that allowed them to live and work legally in the United States until the fourteen-defendant trial came to a close. In exchange, they would have to offer their testimony in court. *(Courtesy of Univision)*

Flowers, letters, crosses, teddy bears, and mementos decorate a chain-link fence on the highway at the spot in Victoria, Texas, where the truck, filled with immigrants, pulled over. *(Courtesy of Univision)*

with him, but there's no way to talk to Toño. Marco had a little cough." That cough, made worse by conditions inside an airless truck trailer, may have become fatal for Marco Antonio.

"Look, I don't want to put you in a compromising position," Carolina said. "If you see him, give him a kiss and a hug, and tell him I love him very much."

"Don't worry," said José Antonio's relative. "Everything's fine." What Carolina didn't know, however, was that as she was speaking on the phone, José Antonio and his son were already on their way to the United States.

On the afternoon of Tuesday, May 6, José Antonio went to the home of his mother, Cristina León Soto, to say goodbye. His mother had been divorced from José Antonio's father, José Luis Villaseñor, for many years. Marco Antonio was with his father on this farewell visit. It was a sweltering afternoon in his grandmother's modest house in the state of Mexico. They all ate together. As they finished up their rice pudding, his favorite dessert, José Antonio confidently turned to his mother and said,

"I'm going."

José Antonio, 31 years old, told his mother of his plans: he wanted to get to Houston, Texas, and then make his way up to Denver, Colorado, where his brother Jesús, also known as Chuchín, was the manager of a Taco Bell restaurant. Jesús, in fact, was the person who had lent him the money to pay the coyotes.

According to one of José Antonio's relatives, José Antonio fell into a "deep depression" after separating from Carolina. It was around then that his 29-year-old brother Jesús spoke to him on the phone from Denver and told him,

"Toño, you have to get out of there. Life goes on." Around February of 2003, just after his separation, José Antonio began making plans for the trip to the United States. His main concern, however, was that his child, Marco Antonio, might eventually get taken away from him, and he did not want to leave for the U.S. without him.

José Antonio's relatives insist that Marco Antonio wanted to be with his father, and that he was constantly saying things like,

"I love papa Toño, I love papa Toño." They also say that Carolina voluntarily gave the child up to his father, and they do not make any mention of the fight that erupted between José Antonio and Carolina in the middle of March.

Carolina states very clearly that she never gave José Antonio permission to take their child to the United States. In any event, Marco Antonio did end up living with his father after his parents' last dispute. According to his relatives, that was what "gave Toño back his soul. He began dreaming about the future again, and even said, 'We're going to learn English, my son.' " José Antonio and his son were very close, they say, and got along famously. Whenever Marco Antonio lost sight of his father, he would call out,

"Where's papa Toño? Where's papa Toño?"

Little by little José Antonio came up with some very convincing reasons to go north. He told his mother and brothers as much:

"I am going to have the opportunity to grow and improve myself and Marco Antonio will go to very good schools. He will find his calling in life." Now that he had Marco Antonio, José

Antonio began to plan the trip to the United States. He sold everything he had: his taxi, the stereo he used for listening to Vicente Fernández music, his furniture. All he had left was a box of photographs and some personal belongings that he left with a neighbor, Doña Socorro. He had been renting his modest house on the Calle Socorro in Mexico City, and so it was relatively easy to leave behind. Carolina, however, did not find out about these preparations until much later.

In April of that year, José Antonio contacted a couple who put him in touch with two other people in Reynosa, Tamaulipas, just across the border from Texas. If he wanted to get from Reynosa to Houston, José Antonio would have to pay them 40,000 Mexican pesos (around $4,000) for himself and 20,000 pesos for the boy. One half would be due after crossing the border, the other half in Houston.

On Wednesday May 7, 2003, José Antonio's 33-year-old brother, José Luis, took José Antonio and Marco Antonio to Central Norte, the bus station serving the area of northern Mexico all the way up to the U.S. border. There, they would board a 6 p.m. bus for Reynosa, close to the border. José Antonio and his son arrived at Reynosa without too much trouble, and everything seems to indicate that their contacts turned up to meet them because father and son did spend the nights of May 8, 9, and 10 in a safe house in Reynosa, Tamaulipas.

On Sunday, May 11, with the help of their contacts, José Antonio and his son crossed the border into the United States, illegally. Exactly how they made the crossing still remains unclear, but it seems that they did it by hiding in a car.

"They took us across in a car," was all José Antonio said over the phone to his 36-year-old sister Emma, or Mimi, the oldest of the family's five children. "Don't worry, we already made it over. I'm in Harlingen, in a safe house. I'm fine. The boy is fine. Don't worry about us. They're going to take us from here to Houston in a tractor-trailer, with air conditioning. Just six of us, it will be easy, comfortable."

That very same weekend, Mimi spoke with her first cousin, Salvador Villaseñor del Villar, and relayed what José Antonio told her.

"She sounded really worried, desperate," Salvador recalls. "Toño's gone," Mimi said to him. "Toño's gone. He didn't have to go. I don't know why he went." In retrospect, Salvador surmised that "she sensed something" wasn't right. In the end, Mimi would be the only person who actually spoke to José Antonio after he crossed the border.

Nobody in the Villaseñor family paid much attention to Mimi's concerns. After all, two of José Antonio's brothers, Jesús and José Luis, had illegally crossed the border into the U.S., and nothing had gone wrong. There was no reason to suspect that it would be any different with José Antonio and his 5-year-old son. But as far as Salvador was concerned, the route his three cousins had taken into the U.S. was still "the death trip."

José Antonio was a robust man who liked soccer and running. Some time before, he had spent a year and eight months in a seminary where he had learned a little bit of Latin and Greek. During his time there, he had seriously considered becoming a priest. His family felt that he had both the physical and spiritual strength to endure whatever obstacles crossed his path in his journey north.

Just like all the other immigrants in the safe house in Harlingen, José Antonio and his son were taken to a clearing in the outskirts of the city on the night of Tuesday, May 13. But if they were expecting to travel to Houston with just four other people in an air-conditioned truck container, they were in for a serious surprise because, in fact, they would be traveling with dozens of other passengers in the back of the truck. Contrary to what José Antonio had told Mimi, things were not going very well at all.

José Luis guesses that his younger brother, José Antonio, initially refused to get into the truck when he saw how many people would be traveling inside, and that he may have been obliged to board the truck with threats and physical force. The other possibility is that, because he was traveling with a small child, José Antonio had been one of the first to arrive and board the truck, and that it wasn't until later, when it was too late, that he realized the trailer might turn into a deathtrap.

Inside the truck, it seems that many of the immigrants were particularly concerned about "the little boy," even though they didn't know his name. Everybody wanted to know what was going on with "the little boy."

José, who was already a grandfather, kept tabs on what was happening with Marco Antonio and remembers watching as the little boy's father lifted him up over his head so he could breathe a little easier.

"When we started to feel the heat," José recalls, "two men broke the truck's tail lights. All we kept saying was, 'All right, let's take turns so the boy doesn't die.' But it was impossible; the heat was too much for him—heat from all the bodies in there and the suffocation from the lack of oxygen."

Israel remembers asking people to step aside so that José Antonio and his son could get to the back of the truck.

"Please, for the sake of the child, get out of the way, let the father take the boy to the hole so he can get some fresh air," he told them. But Marco Antonio was already taking in his last breaths. "The people in the back of that truck were desperate, all they wanted was a little bit of air. Everyone was completely frantic."

Enrique had already told José Antonio to bring his son over to the holes where a tiny bit of air was steaming in. Even so, it was no easy task.

"They brought the little boy over, so he could breathe in some fresh air," Alberto remembers, "but we had to practically stick our noses out the window just to even get a little air, because we were at the back end of the truck."

Alberto had spoken to the boy a little, back in the safe house in Harlingen. At that time, he only knew him as "Antonio," the boy's second name. He remembers Marco Antonio playing in the house and being very convivial with everyone. Even at the tender age of five, the little boy seemed to have a way with people.

Alberto recalls that once everyone was inside the truck, Marco Antonio's wails were the first sign that something was very wrong. What he remembers most vividly, however, is José Antonio's anguish. "The little boy's father cried out that we had to stop the truck, but there was no way," Alberto said.

It was impossible.

"The boy died, he died!" José heard.

"Oh my God," he squeaked out, his voice reduced to a whisper, and then hung his head against his chest.

6

THE DEAD AND
THE SURVIVORS

As soon as the trailer doors were opened, many of the passengers began jumping out. Enrique was the fourth or fifth to step down.

"When I turned around, you couldn't see anything, it was all dark. Right at the door I managed to see two or three people. I thought: they're passed out, the ambulance will come soon, they'll give them some oxygen and then they'll feel better. When I got down from the truck, I stepped aside. I wanted to see who had been driving, because the person who opened the door for us was not the driver. It was someone else." Enrique was not certain whether the driver's companion had been a man or a woman, but he was able to catch a glimpse of two people heading out in the driver's cab.

Enrique, meanwhile, was still trying to figure out exactly what had been going on for the past four hours. He had made it out alive. Now, as his breathing very slowly normalized, his primary concern was that he might get caught by immigration agents.

The truck had pulled over on a tiny rural road in the outskirts of Victoria, Texas, next to the Hilltop Exxon Truck Stop, close to U.S. Highway 77. Enrique was extremely thirsty, and off in the distance he could see something that seemed to be shining, like a water fountain. Before running toward it, he went back inside the trailer. This time, with the moonlight above, he was able to see things a bit more clearly. And he now realized that a number of the people inside the trailer were dead. Taking another sip of water, Enrique turned away from the trailer for good. He knew that the immigration agents would throw him in jail if they caught him. The judge's warning was still crystal-clear in his mind. And so the next thing he did was turn and walk into the open field, without water, without food, without even a shirt.

"I was hidden by the path," Enrique continued. "But along the way two guys from El Salvador joined me, and the three of us walked together. One of them was underage—less than 15 years old, I think. It was their first time in the United States." At some point, the three men found themselves walking across land that was part of a private ranch. But then, suddenly, the Salvadoran boy fell flat on his back, in an area where the grass was two feet high.

"He's going to suffocate in that hot grass," Enrique thought. "He's going to suffocate." Enrique lifted him up, took out the water he had brought with him, and sprinkled some on the young man's face. Once he was feeling refreshed, the three men continued walking up toward a hill.

Enrique and the two Salvadoran men hid out for the entire day of Wednesday, May 14, and all morning on Thursday, May 15. But by three in the afternoon on Thursday, the young Salvadoran man couldn't take it any more.

"I'm hungry, I'm thirsty," he said to Enrique two or three times. Enrique, the leader of the group, began to think that they might die if they didn't eat or drink something. They kept on walking until finally they reached the main house on the ranch. Enrique didn't want to get any closer. He wasn't wearing a shirt, and he didn't want to frighten the owners. The 15-year-old boy, who did have a shirt on, went up to the house to ask for a little bit of water.

"The owners of the house came out," Enrique said. "They were Americans. They called us over and gave us water, bananas, and some bags of potato chips. The boys from El Salvador didn't understand any English. So I started talking to the man, and the woman said:

'I'm going to make you something to eat, I'll be right back.' And she left. When she came back she brought us lunch, and the man began telling us that eighteen people had died [in the truck]. I was surprised—I thought only the little boy had died, I didn't think any adults had gone, too. That day, right there, I found out, two days later, that there were eighteen dead and six or seven in critical condition in the hospital. [The gentleman] asked us if we had been riding in that trailer and I said yes. I was answering and translating for the other men so that they would know what was going on. Then he told us that he would be glad to help us out but to him, of course, 'helping us out' meant turning us over to the authorities, to the police, whom he had already called."

When the police arrived at the ranch, someone speaking through a loudspeaker ordered Enrique and the other men not to run. Enrique complied, and the police did not handcuff them: they simply ushered the men into the patrol car and drove Enrique and the two Salvadorans back to the spot where the

trailer had stopped. Then, they turned them over to the immigration service. They had to wait for twenty minutes; during this
time, the police bought them a few large bottles of water.

Alberto, just like Enrique and the Salvadorans, also tried to
make a run for it after he stepped down from the truck.

"There were so many bodies scattered around, but I figured
that most of them were just passed out. I remember touching
my friend's uncle. I touched him and said to everyone, 'He's still
alive.' I touched him, and I could tell he was still alive. But later
on they lowered him down, started to treat him, and by then,
no." By then he was dead.

Alberto, despite the terrible shock of what he had just witnessed, did not stay where he was.

"I tried to run, I tried to run toward my dreams," he says.
"I remember, I jumped over a fence, and when I saw some of
the other passengers lying on the ground, and I called out to
them, 'Come on, come on, let's keep walking, maybe we'll find
someone who can give us a ride to Houston.' A small group of
us—lots of young Salvadoran women, the majority, and some
Hondurans—started off, and we walked, we walked for a long
time."

At dawn on Wednesday, May 14, this group of eight immigrants started walking away from the truck in the hopes of
averting the border patrol authorities. One of the people in the
group was Matías, who had a cell phone with him. He was the
person who had called 911 from inside the trailer—the call that
had been answered by an operator who did not speak Spanish.

"I didn't meet Matías until after we got out of the truck," Alberto recalls. "As we began to walk he suddenly said, 'I have a cell phone.' This seemed strange to me. Why did he have a cell phone?" Alberto was worried that they would be in for trouble if the immigration service agents picked them up and saw that someone in the group had a cell phone.

"I think you ought to throw that phone away," Alberto said to Matías, after explaining his reasons.

"Yes," Matías replied, simple and straightforward. But he didn't throw it away.

"We kept on walking," Alberto continued. "We were with some women, I think they were from Honduras. Some of them had lost their shoes, and we wrapped our shirts around their feet so that they could keep walking. We walked until we reached the first river and then we drank some water." Everyone was exhausted, so they lay down to rest.

"Who's willing to go up to the road to see if we can get a ride from someone?" Matías asked.

"I'll go with you," Alberto said to him, even though he didn't know him very well.

"Everyone else seemed exhausted, but I was all right," Alberto continued. "I walked with him up to Interstate 77, but all we saw were ambulances going by. We both got scared when we saw that. And then there were patrol cars, policemen."

"We'd better go back," Alberto ventured.

"Yes," Matías agreed. "This isn't looking too good."

When they went back to the spot where they had left the rest of the group, nobody was there. They had all fled. The situation was getting more and more complicated for Alberto and

Matías. The next thing they knew, a voice in Spanish, speaking over a loudspeaker, was asking them to turn themselves over to the police. Both men decided not to.

"We're better off sleeping here for the night," Alberto said to Matías. It was four in the morning, two hours since they'd stepped down from the truck.

"As soon as the sun starts to come up, we'll go," Alberto continued. But they didn't have to wait that long. The cold air woke them up. They were shivering.

"It's getting late," Matías said, and they started walking. Matías hadn't abandoned his cell phone and had in fact made another call to his aunt and mother who, right then, were in a car somewhere near Victoria, Texas. They told Matías that they were at a rest stop on the side of the road, but since neither he nor Alberto knew where they were, they kept walking— in the opposite direction of where they had left the truck. Whenever helicopters flew overhead, they hid among the trees and bushes.

"I think we're going in the right direction," Alberto very optimistically said to Matías. This was Matías's first trip to the United States, and even though they had only met a few hours earlier, Alberto felt responsible for him, in a way. As they continued walking along the side of the road, Matías kept trying to call his relatives. According to Alberto, Matías made at least ten phone calls.

"With our luck my battery will run out," Matías said grimly.

As the sun went up, so did the danger of being caught by the police or immigration agents.

"I think we're better off on the other side," Alberto said to Matias, noting that the trees on the other side of the road offered a better hiding spot. But just as they were crossing the road, they were spotted by a group of journalists in a white truck with a television antenna. They honked the horn, but the two men kept on walking.

"I think they were the ones who called the immigration agents," Alberto recalls. "We were still walking—or running, more like it—and finally we got to a river, real wide. We had no other choice but to get in and swim."

"I'm not too good at swimming," Alberto said, eyeing the water level.

"Well I'm probably worse," Matías replied, still with a sense of humor.

The two men moved closer to the river, to get a drink of water. They were squatting down when suddenly an immigration officer caught them by surprise.

"I spun around and saw the immigration guy, real big and blond," recalled Alberto, suddenly transported back to the moment he was caught. "Damn!" I said. "They got us."

"I've got plenty of help for you," the agent said, insisting that they stop drinking the water from the river. "I've got water in the van." Without a fight, they followed the agent back to his vehicle.

"It was just a trick to get us into the van," Alberto recalls. "He didn't have any water." It was only as they traveled back down the road that they realized how very far they had walked. But now Alberto and Matías were in for another shock, for they were being driven back to the spot where the trailer still sat, alongside the dead bodies of many of their fellow passengers. Until that moment, neither of the two men had realized that so many people had died.

"I wanted to think they weren't dead; I just couldn't get it through my head. That was the saddest moment of my life," Alberto remembered. "The moment I turned around and saw those sheets, and that someone so young was wrapped in one of them. You could see his face because the sheet kept flying up from the wind. I saw all those bodies lying all over the place, and I felt so terrible. I still don't know why they brought us there—only they know why they did it. It was like some kind of punishment for us. Those bodies, those things I saw there . . . that's what I dream about sometimes."

"I've got water, man," Israel said to José. "Look, drink." José was going in and out of consciousness. When people had started getting off the truck, he hadn't even realized.

"Help me, son, I can't get down," José had said to his nephew. After stepping down, José whipped around to look inside the truck container and shouted,

"Tito, we're here. Get down. What are you doing there?" But Tito, which is what they all called his brother-in-law Hector, did not get down from the truck.

José thought that Tito had passed out, like he had just before. But he didn't have the strength to go in and look for him, so he took a sip of water from one of the bottles that Israel had brought him.

"Water, we need water," the other immigrants pleaded, and José passed one of the two water bottles to the people inside the truck. Then he walked over to the gas station convenience store, but they had run out of water by then.

When he returned to the truck, his nephew Israel was looking for his uncles Roberto and Serafin, as well as for his friend Hector.

"My uncle, my uncle, I found Uncle Serafin," Israel shouted to José.

"Where?" José asked.

"Come here, and help me get him out," Israel shouted back, indicating the inside of the trailer.

It wasn't easy getting Serafin down from the truck.

"We almost dropped him right there," José remembered. Finally, they set him down on the ground, face up. That was when they realized he wasn't moving.

"Give him air," José said to Israel. Without any kind of first-aid training, Israel began to blow into his mouth, in a failed attempt at mouth-to-mouth resuscitation. As he did this, José pressed down on Serafin's chest to try and revive him.

"There was no way," Israel recalled. "By then he had foam coming out of his mouth."

When he realized that Serafin was dead, Israel left him lying on the ground and went back into the truck to find his other uncle, Roberto, and Hector.

"Uncle Roberto, Uncle Roberto," he shouted, but no one responded. Inside the trailer everything was still dark; dawn was several hours away. Israel had to move seven dead bodies out of the way before he found the bodies of Roberto and Hector.

"Thank God I am alive," Israel thought. "God wanted to take them home, and he will forgive them." There was nothing else to do. He didn't even try to drag them out of the truck.

"There's nobody, there's nobody," Israel said to his uncle José over and over again. What he really meant was that there was nobody alive.

"Yeah. So what do we do?" José said.

"Leave them there." They took the personal identification off their dead and decided it was time to head for the hills. They wanted to get as far away from the tragedy as they could, and with a little luck, they might be able to keep clear of the immigration police, too.

"Uncle José, wait up," Israel shouted to José, who had already started walking away.

"Come on!" José called out in response. They were both very shaken by the death of their relatives, but more than anything they were tired, thirsty, and weighed down by their sweat-drenched clothes.

"We were really soaked," José remembered. "Our clothes were heavy from sweat, pure sweat." They didn't get very far. About 300 meters from the truck they saw a house, and they sat down for a rest in the front yard, but the grass was thorny. There was no way they would be able to sleep. So they just sat and waited until dawn. It was foggy, and José sensed that it would rain soon.

"Don't move!" they suddenly heard. The voice came from a patrol car that had just pulled up at the house. "We're not going to do anything. We just want to help you." José figures that it must have been around seven in the morning on Thursday, May 14, when the patrol car discovered him and Israel. The police officers gave them some bottles of water.

"I was dying of thirst by then," José remembers. "I rinsed my mouth out and then downed the whole bottle. After that they put us in [the patrol car] and took us to Victoria, to the hospital.

And then they told us, 'We're going to wash your clothes,' because we stank from all the sweat."

José was feeling very queasy. After they gave him back his clothing, freshly washed, he told one of the doctors at the hospital that he was a diabetic, and they checked his sugar levels.

"You're very high," one of the doctors said. "Your sugar." They immediately gave him a shot of insulin. From the hospital they were then taken to the Victoria community center, near the fire station. There, José and Israel met Carolina Zaragoza, the Mexican consul in Houston in charge of civil protection. Sooner or later, all the immigrants from the truck ended up at the community center. After taking down their information, Zaragoza informed them that they would be allowed to make one telephone call each to Mexico. But it was a call that neither José nor Israel felt much like making.

"Listen." That was the first word José said to his wife, María Ramírez, after dialing the number of his house in Pozos, Guanajuato.

"Who's calling?" she asked.

"Me," said José, drained of all energy.

"Who?" María asked again.

"Reyes," he replied, using his second name.

"Oh, honey! Where are you?"

"I'm here, in Victoria."

"You came back already? Why?" José's wife had assumed he meant Victoria, Guanajuato, not Victoria, Texas.

"No," José said, and then clarified it for her: "There's a Victoria here, too."

"Should I send you the money?" she asked, thinking they would need it in Texas to pay off the coyotes.

"No, no. Don't send any money." That was all he said.

"Oh," said María, to fill the silence. Then she went on to tell him about something she had just seen on television. "There was this tractor trailer, a really awful tragedy. Do you know what happened?"

"Well . . ." José replied, summoning up his strength. "You know what? We were inside that tractor-trailer."

"*Ay* . . ." she said, fearing the worst. "What happened?"

"Well, something . . ." he said. "Serafín died in there. So did his brother and, um, your brother [Hector]."

"No, no, no, no . . . !" she cried at the top of her lungs.

"Yes. I know. It's awful," José continued, trying to console her. "But we have to accept it, because that was God's will."

María began to cry uncontrollably over the loss of her brother Hector. But José soon had to hang up the phone. There were other people waiting to call home to Mexico.

Sheriff Henry García Castillo, of Victoria County, Texas, was one of the first to arrive. Just after three in the morning on Wednesday, May 14, 2003, he parked his patrol car close to the scene of the crime. The bodies of the dead were still hot. They were all hot; one news report suggests that some of the survivors had fevers of 105 degrees when first found.

Sheriff García Castillo climbed into the truck container. Slowly, methodically, he examined each of the bodies to check if anybody was still alive. Thirteen, fourteen, fifteen motionless bodies.

"This is an outrage," he thought. "Nobody should have to

die this way." He continued making his way through the dead bodies, until he came across a child. "It broke my heart. What can I say? From what I understand, the man lying next to him was his father." Sixteen, seventeen dead.

After making sure that there were no more survivors, the sheriff began looking for evidence, anything that might help explain what had happened inside the trailer. While he was doing this, authorities launched a search, which would last fifteen hours, for the immigrants who had survived the tragedy and who were now scattered around the mountains of Victoria County. Sheriff García Castillo's main concern was that these people were likely to be ill, thirsty, hungry, and in need of emergency medical attention. The search ended at six in the afternoon.

Sheriff García Castillo does remember one incident that was a bright spot during that terrible day. One of the girls who had survived the trip had been taken to the Victoria community center. When the people there realized it was her birthday they bought her a cake.

"We all have a little heart," he surmised. "People wanted to help out, that's all."

It was about three in the morning by the time Karla Chavez and Freddy García reached the spot where the tractor trailer had pulled over in Victoria, Texas. They couldn't see much of anything. The two doors at the back of the truck trailer were open, and they could just barely see the bodies lying around, both inside and outside of the truck.

According to Karla, she and Freddy were in Corpus Christi, Texas, when they first received a phone call, ostensibly placed by Ema, Víctor Rodríguez's wife, alerting them that the truck had stopped en route. The Rodríguez family was concerned; they had sent eleven immigrants on that truck.

As soon as they heard the news, Karla and Freddy went to Victoria. But by the time they arrived, there wasn't much for them to do, so they decided to go to a nearby hotel. A receipt from a local hotel confirms that Karla spent the night of May 14 in Victoria.

Norma Gonzalez, however, another defendant in this case, later told an undercover agent that Karla had been traveling behind the truck for practically the entire ride.

Who was telling the truth? Had Karla really been in Corpus Christi when all this happened, or had she been trailing the truck for those four hours before they finally opened the doors?

Driver Tyrone Williams didn't want to see any more of this macabre spectacle, so he disconnected the trailer from the driver's cab (which bore the New York license plate number 47525-PA), and he fled the scene. Williams was out of control; he didn't know what to do. He had never imagined that he would find so many immigrants in the back of his truck, or that any of them would die before the ride ended. His decision to flee the scene of the crime was a determining factor that, later on, prosecutors would cite when they announced they would be seeking the death penalty for him—and none of the other defendants.

Without thinking much, Williams drove all the way to Houston, Texas, and went directly to the emergency room at the Twelve Oaks Medical Center. It was 5:15 in the morning. In her medical report, the nurse who treated him wrote that Williams had been extremely distraught and anxious, and that he had told them a very strange story. Williams had told the nurse that he had been driving a tractor trailer that he had believed to be empty until, suddenly, the driver of another truck on the highway signaled to him that some of the lights on his trailer weren't working. When he stopped and opened the back part of his truck, he realized that it was filled with undocumented immigrants. He got scared, he said, and disconnected the truck container from the driver's cab. And then he drove to Houston. The hospital informed the police of the incident at approximately six in the morning. The call was urgent: in Victoria, Texas, there were a number of undocumented immigrants who might be on the verge of dying.

At 7:20 in the morning, Agents Thomas Cason, of the U. S. Immigration Customs and Enforcement (ICE) and Gary Thomas, of the FBI, arrived at the Twelve Oaks Medical Center to question Williams. There, Williams admitted that he was the owner of the trailer in Victoria, Texas, and that he had suffered a panic attack when he saw some fifty immigrants inside the trailer. The agents read Williams his right to remain silent and to seek the advice of an attorney, and then they arrested him. Shortly thereafter, Williams was taken to the immigration service's Houston office for further questioning. According to the official case documents, at ten forty-five in the morning on Wednesday, May 14, Williams gave up his right to an attorney and told the ICE agents John Chiue, Minh Lu, and Supervisor John Swallack what had happened.

Eduardo Ibarrola, the consul general of Mexico in Houston, was not accustomed to receiving phone calls from the Mexican Embassy in Washington before breakfast. For that reason, when the embassy placed an urgent call to him on the morning of Wednesday, May 14, he had a feeling that it would be one of the longest, most difficult days of his career as consul general. The early reports seemed to indicate that an accident had occurred; that was all the information they had in Washington. But by the second phone call, it had become eminently clear that they were facing a tragedy of staggering proportions.

Consul Ibarrola immediately telephoned Carolina Zaragoza and Carlos García, the two people responsible for providing consular protection to Mexicans outside their home country, and told them to get ready to go with him to Victoria. Ibarrola, Zaragoza, and García left Houston at 9:30 a.m. Two hours later they arrived at the spot where the truck was parked. Despite the information they had already received, none of them was prepared for what they would see when they got out of their car.

"We saw the bodies of the people who died," Consul Ibarrola remembered. "Some of them were inside the truck, others had been taken out already. Some of them had no clothes on. You could see that the heat had been unbearable—that was why they had taken off their shirts. What I saw that morning was one of the most horrifying things I have ever had to witness in all my professional life. Especially the body of that little boy, who was 5 years old. That was one of the saddest things—

perhaps the saddest—that I have ever had to see. From what the survivors told us, the little boy was the first to die. And the father, unfortunately, he died as well."

The U.S. authorities were in charge of investigating the incident, but consuls Ibarrola, Zaragoza, and García immediately began talking to the survivors to hear their version of the facts. The consular file, which they would later use to write a report for the Mexican government, includes statements made by various Mexican immigrants who say that they had climbed into the truck container as quickly as possible so as to avoid being caught by the *migra,* the U.S. immigration authorities.

"What they tell us," Ibarrola said, referring to the consular file opened that same day, "is that from the beginning, [the immigrants] knew that there were too many people inside the trailer, and that once inside, they felt a tremendous level of heat from the very start. They were traveling in complete darkness. They were so frantic and so desperate from the mounting heat that they managed to break two of the truck's tail lights and carve out a couple of holes which allowed them to breathe in a little oxygen through the open spaces from the broken tail lights. I don't think any drivers on the road heard their cries. They also say that they began waving items of clothing— T-shirts, shirts—to get the attention of the people driving behind them on the road. The tail lights, however, were dangling down from a cable, and it seems that some other drivers on the road noticed this and signaled to the truck driver that something was going on inside the trailer. Apparently, that was what made him stop his truck in Victoria, about halfway through the ride. That, in general terms, was what the immigrants told me."

Sixteen of the nineteen people who died and thirty-two of

the fifty-four survivors turned out to be Mexicans. The majority
of these Mexicans were from the states of Guanajuato (inciden-
tally, the home state of Mexican president Vicente Fox), Puebla,
Tamaulipas, San Luis Potosí, Guerrero, Mexico State, Nuevo
León, Veracruz, and the Federal District. Very few of them had
ever even met before arriving in the U.S. Most of them first laid
eyes on each other in the safe houses in Harlingen, Texas, run by
the coyotes. On average, they paid between $1,000 and $1,500
each to get from Mexico to Houston.

The representatives of the Mexican government made a pre-
liminary, tentative list of the Mexican immigrants who had survived. The overwhelming majority of the fifty-four survivors
were in fact Mexican. After receiving medical attention, they
were taken to a community center in Victoria. Ibarrola feels that
there was, "from the start, an excellent level of cooperation between the U.S. authorities, the authorities in Victoria city and
county, the U.S. Attorney General in Houston and the Homeland Security authorities."

In the aftermath of the tragedy, the authorities of both countries seemed to be working in perfect coordination with one another. But it was still too late. The coordination may have been
perfect after the tragedy, but it hadn't been good enough to prevent it from happening.

"They were all migrant workers," said Consul General Ibarrola. "They were all headed for Houston to earn money. They
all wanted to help their families." The thirty-two Mexicans were
released on bail, which the Mexican government paid on their
behalf. One of the thirty-two survivors was 16 years old, and
because he was a minor he was sent into the custody of his

family in the United States. The rest of the survivors were granted work visas and Social Security numbers so that they could remain in the United States for the duration of the trial against the traffickers.

On the morning of Wednesday, May 14, Carolina Acuña woke up in her Mexico City home to give her father his breakfast. Her mother was away, visiting some relatives in Chiapas. Just as she did every day, she turned the television on to Channel 13, and on Televisión Azteca's morning news program she learned that a number of immigrants had died of asphyxiation in the back of a tractor-trailer. At the time, Carolina didn't think much of it.

"I never would have imagined that my son was one of them." The early news reports did not offer the names of any victims. As far as the Acuña family was concerned, there was no reason for alarm. After breakfast, Carolina's father went out, leaving her alone in the house.

In the middle of the morning, to her surprise, Carolina received a phone call from Virginia, one of José Antonio's two sisters, and also Marco Antonio's godmother.

"You heard, didn't you?" Virginia asked her. "Marco and Toño died."

"I can't believe the sonofabitch!" replied Carolina, still not quite realizing what had happened. She was used to José Antonio and his threats, but she never dreamed that they would culminate in the death of both father and son. At that moment, Carolina still had not made the connection between José

Antonio and Marco Antonio and the news report she had seen earlier that morning, the one about the immigrants who had died from asphyxiation in the truck.

"Caro, I didn't call you up to get into a fight," said Virginia, in a conciliatory tone.

"Please, can't you understand my pain, as a mother?" Carolina said.

They exchanged a few more words; Carolina listened as Virginia told her that José Antonio and Marco Antonio had died in the truck incident. She doesn't remember which of them hung up first but she does remember that, as she replaced the telephone receiver in its cradle, she clung to the window in her mother's bedroom and said,

"It's not my son. It's not my son. No. No. No . . ."

Soon afterwards, her brother Natalio arrived at the house and, without saying a word, hugged her tight.

"It can't be true, no, no . . ." Carolina whispered between her sobs, unable to accept the fact that her son was dead.

"Let's just wait a little while," Natalio said, trying to console her. He was the first to see the list of names appear on the television screen, and it was true: José Antonio Villaseñor and Marco Antonio Villaseñor had died.

Carolina felt something akin to an electric shock when the first photograph of her son greeted her from the television screen. One photo showed him on vacation in Veracruz, while another showed him just after he had earned his kindergarten diploma. Carolina didn't know who had given the TV those photographs: Doña Socorro, the neighbor with whom José Antonio had left a box of mementos; Doña Refugio, his landlady; or his own family. But there it was: the image of her son, on television.

Carolina then received a phone call from the Mexican Ministry of Foreign Affairs, but she asked one of her sisters, Socorro, to take the call so that they could find out what they had to do in order to identify and claim the bodies. Even then, Carolina didn't fully believe that her son was dead. The TV news program broadcast a list of the people who had perished, but all it said was "José Antonio, 38 years old" and "Marco Antonio, 8 years old." There was still a shred of hope, because there was some discrepancy in the ages cited: José Antonio was only 31, and Marco Antonio, 5.

"Any little bit of hope is enough," Carolina said to herself. "I felt so much rage, such impotence. I wanted to scream out at the world, at someone, anyone, please have pity on me! Anyone who could tell me that this wasn't true, no, it's not true, this is not true. I was so scared, so furious. And then I said: 'No, not until I see it for myself.' " Carolina thought that perhaps they had been mugged or attacked and hadn't gotten into the truck. She knew, for example, that José Antonio always wore a silver bracelet with his name on it. He never took it off. Carolina needed to see something tangible that would prove to her that José Antonio and her son were really dead.

"It isn't Marco," she insisted. "Anything but that. This can't be happening to me."

"The boy is in here," said one of the policemen guarding the trailer to Mexican Consul Carolina Zaragoza. Incredulous, she turned around to exchange a glance with the other representatives of the Mexican consulate in Houston who had joined her on the trip to Victoria. Nobody had told them, in any of their telephone conversations, that one of the victims was a little boy.

"It was a terrible blow. Everyone was devastated when they saw the little boy with his father," Zaragoza recalls. "They were right next to each other, the father's body protecting the boy." In her twenty years as a career diplomat, Carolina Zaragoza had never witnessed anything quite like this. "When you see something like that, all you can feel is a terrible sense of impotence. All the little boy had on was his underwear. There were others, too, half naked, without shirts, or T-shirts."

Seeing Marco Antonio made Carolina think of her own son. Walking away from the trailer for a moment, she breathed deeply and thought to herself,

"My brain has to freeze right now. We have to get to work." She seemed to automatically respond to her command, and she felt a tremendous surge of energy rise up inside of her, energy that she would no doubt need if she were to make it through the next few days.

By now it was midday in Victoria, Texas, and some twenty news organizations were already covering the tragic news. Wasting no time, Consul Zaragoza began counting the number of dead bodies inside the truck container: one, two, four, eight, twelve, fifteen, seventeen. Several of the bodies were already covered in white sheets. There was nothing more they could do in there.

Carolina and the other representatives from the Mexican

consulate then went to the Victoria community center, where they met with the majority of the survivors, and immediately began interviewing them, taking down names and home towns. With a couple of cell phones from the consulate, they helped the survivors get in touch with their family members back in Mexico. It was around one in the afternoon. In Mexico, the radio news programs had already begun to cover the catastrophe that had befallen a group of undocumented immigrants, but the information was still scant.

"The first thing the survivors wanted was to get in touch with their relatives, to tell them that they were fine, that they hadn't died," Carolina remembered. Her primary responsibility at that time was to identify the survivors who were Mexican citizens. But suddenly, a Honduran woman approached her, frantic.

"Help me," she told the Mexican consul, producing a tiny slip of paper, perfectly folded, from her pants pocket. On it was the telephone number of the people who had taken her son across the border from Mexico. "My son, my 3-year-old son stayed back. Please help me, I have to find my son."

Carolina immediately understood what had happened. The coyotes had suspected that the 3-year-old child would not be able to endure the ride in the trailer and decided to transport him across the border separately from the rest. Once they all arrived in Houston, they would deliver him to his mother. But now that things had taken such a terrible turn, the mother was terrified that they would disappear with her son and she would never see him again.

Carolina worked fast: first she spoke with a group of immigration agents and showed them the slip of paper with the telephone number of the people who apparently had the Honduran woman's child. Rather unwittingly, her efforts resulted in a massive search operation that swept through all of southern Texas.

Once they had completed the full list of survivors at the community center, Consul Zaragoza departed for the local hospital where she met with another group of women, this time from El Salvador and Honduras.

"I chatted with them. They were young women. They told me that things had gotten really desperate inside the trailer. They told me how they lifted the little boy up so that he could breathe. Some other people from El Salvador told me the same thing, that they had helped lift the boy up, too."

But their efforts were all in vain.

Carolina believes she understands why all the mortal victims were men. Why didn't a single woman die?

"The men grew more panicked, maybe. They went from one side of the trailer to the other. The women, on the other hand, took it differently, they just said, 'I'm better off staying put,' and that way, they didn't waste any unnecessary energy. That was what saved them."

While Carolina chatted with the Central American group at the hospital, she couldn't help but reflect on how very lucky they had been.

"The temperatures that month, and that day especially, were so high," Carolina Zaragoza recalled. "Forty more minutes in that trailer, and all of them would have died." But she didn't want to tell them that. They had already been through enough.

* * *

"No, no, no, they never would have survived." That is what nurse Gilda Miller believes, had the immigrants remained trapped in the trailer for much longer.

In her twenty-six years as a nurse, Gilda Miller had never seen anything like this. Miller works in the pediatrics department of Citizens Medical Center, very close to where the immigrants died. Sometime around noon on Wednesday, May 14, a number of young women began to turn up at the hospital.

"The girls were all very emotional, crying as they described what they had been through. They told me that what scared them the most was that each time they took a breath, it might be their last.

"They heard lots of things, arguments, people crying." The young women told the nurse that even though they couldn't see each other, they somehow managed to stick together inside the trailer, and they prayed that someone would come and help them.

"People suffering from a lack of oxygen tend to get very belligerent. They start fighting. They don't behave normally . . . until someone faints. A person who isn't receiving enough oxygen or water will normally lose consciousness. That's what happened to many of the people suffering [inside the trailer]."

At the hospital, the group of women were given plenty of liquids and food. The most dehydrated of them were put on IVs.

How long would the truck have taken to get from Victoria to Houston? the Central American women asked Gilda, after she had given them first aid. An hour, an hour and a half, maybe.

They turned to stare at each other. If the driver had gone straight to Houston without stopping, all of them would have died.

Two immigrants, one from El Salvador and another from Honduras, were removed from the trailer in critical condition and taken to the hospital in Victoria. Dehydrated, their lungs barely moving, they were at a high risk for heart attacks. Despite the medical attention they received at the emergency units, neither of them survived. When they passed away, the death count rose to a total of nineteen.

The Wednesday after the tragedy, at around three in the afternoon, María Ortega saw on TV that a group of immigrants had died inside a truck container. She had been waiting to hear from her brother Enrique ever since he had left Mexico for Texas a few days earlier.

The previous night, she had spoken to the coyote who had helped Enrique across the border from Mexico, to ask him how he was planning to get her brother up to Houston. During that conversation, María was told that her brother would be traveling in a small group of eight people, five men and three women, inside a tractor-trailer. The coyote, however, failed to mention that María's brother would be traveling in the back end of the truck.

María grabbed the telephone and called the coyote she had hired to get her brother across the border. They hadn't spoken much before, just enough to agree that her brother would arrive at Matamoros, Mexico on May 10th, and from there he would be led across the border into Texas. And then they had spoken the previous night. But they had never met in person. When she got through to him, María asked the coyote if her brother Enrique had been traveling inside that trailer.

"No," the trafficker told her. Enrique hadn't been in that truck. But something in his voice made María doubt him.

An hour later, at four in the afternoon, María dialed the same number, and once again asked the coyote if her brother Enrique had really not been inside that truck. This time, the answer was different.

"Unfortunately, yes, he was in the truck," he responded, but he didn't have any more information than the news programs on TV. María wanted to know what her brother had been wearing and if he had been carrying a false ID card with a false name. But she was unable to find out anything else.

María and her husband, who lived in south Texas, immediately went to the city of Victoria to search for her brother. They were desperately afraid that he might have died, but they also knew that Enrique was a strong, determined young man. Maybe he had survived the trip and was hiding out in the mountains. All along the roads near the spot where the trailer had been parked and abandoned, they searched for him until nightfall. They didn't find him, but they weren't ready to give him up for dead yet.

The next day, María and her husband went to the Mexican consulate in Houston in the hopes of obtaining some information and brought with them a photograph of Enrique, which they left there. Then they went home, steeling themselves for the worst. But the following afternoon, they received a phone call from the Mexican consulate with the news they so fervently hoped to hear.

"Your brother is fine," María heard over the telephone line. Enrique was alive. But she still had her doubts. What if someone had stolen Enrique's identification? Or if his ID had ended up with someone else's things?

"How can I be sure it's my brother?" she asked.

"It's your brother," they assured her. They went on to describe his appearance until María was sufficiently convinced.

"Yes, yes, that's him!"

At that moment, however, they could not put Enrique on the phone. In fact, María would not be able to see her brother for almost an entire month, because he had been placed in custody to help with the investigation. But María didn't care about that. She knew Enrique was alive; that was enough for her.

Wednesday, May 14, should have been a good day for Norma, the manager of the Houston restaurant, and her employee, whose brother was scheduled to arrive in Houston that day thanks to Norma's help, her contacts, and a payment of $1,800.

But when the employee arrived at the restaurant that day, Norma told her the bad news: the tractor-trailer that had been hired to take her brother from Harlingen to Houston had been stopped by the *migra,* the immigration police. The restaurant's television set was on. As the two women talked about what had happened, a local Spanish-language news program came on the air and reported that several immigrants had died inside a trailer in Victoria, Texas. Upon hearing this, Norma tried to calm her employee, reassuring her that her brother had been traveling in a different tractor-trailer. Hours later, Norma would be forced to admit to the employee that her brother had indeed been traveling in the same tractor-trailer that had appeared on the TV news. What could she do? How could she find out if her brother was one of the survivors?

The following morning, Norma offered to take her employee to Victoria to see if they could get more information. When they got to Victoria, however, they were told there was no list of the people who died, and that nobody had been able to identify the bodies. They decided to go back to Houston and wait for more news on TV. But the names of the victims were not mentioned on the TV news that night, either.

The next morning, Norma's employee went to the Mexican consulate in Houston with a few photographs of her brother. There, a copy was made of one of her photographs, and it was sent to the medical examiner's office in Victoria, where her worst fears were confirmed. Yes, one of the dead bodies had been identified as her brother.

Distraught, she went directly to the restaurant to tell Norma that her brother had died. But Norma wasn't there: she had gone to Monterrey, Nuevo León, Mexico. The restaurant cashier, however, had been instructed to return the $900 the employee had paid for the first installment of her brother's journey. She took the money and left, deeply wounded by her brother's death as well as Norma's failure to face her. She tried telephoning Norma several times, but Norma never answered the phone.

That was when the employee decided she would cooperate with the agents from ICE, the U.S.'s new immigration service.

Norma eventually contacted the employee by phone and offered to help her in any way possible, even financially, to make up for the loss of her brother. There was nothing she could do to bring her brother back; she knew that. But now, Norma's employee wanted justice. The two women agreed to meet at the restaurant on May 20th, at 1:30 in the afternoon. The employee, however, didn't show up alone. She arrived at the restaurant with an undercover ICE agent whom she introduced as her cousin.

During their conversation, Norma tried to convince the employee that she was in no way responsible for what had happened and claimed that she only worked for someone named Karla. That was all. Norma, however, told the employee that she would be willing to pay her money—something "fair"—to give to the dead man's wife or mother, simply as a way of helping them. The employee told Norma that she would have to speak with her relatives about the amount of money, and then she left the restaurant with her "cousin."

Legally speaking, Norma's luck had run out.

Josefina Gonzalez did not learn that her granddaughter Fabiola had been in the fateful trailer until six days after the tragedy. She suspected something had gone wrong, but she didn't know exactly what.

On May 14, one day after having paid Norma Gonzalez $950, Josefina received a phone call from Norma, who told her that the immigration service had picked up the group with whom her granddaughter Fabiola had been traveling, and that several of the immigrants in the group had managed to escape. Norma did not, however, mention that Fabiola had been inside the ill-fated truck trailer.

On Thursday, May 15, Josefina received another phone call from Norma, who informed her that she would not be able to reunite Josefina with her granddaughter, and that she would like to return the $950 Josefina had paid. When the two women met at a Houston gas station later that day, they barely exchanged a word. Norma did not tell her where Fabiola was, and Josefina did not ask any questions.

Josefina had no idea where her granddaughter was.

On May 20, a ICE agent finally called Josefina to tell her that her granddaughter had indeed been inside the trailer where several immigrants had perished. But Fabiola had been one of the lucky ones. No, she hadn't died. A ICE employee had questioned her, and aside from the trauma she had endured, Fabiola was fine.

What really happened inside that trailer?

On Thursday, May 15, 2003, María Henao, a producer at Univision in Miami whose job was to assign reporters to stories, received an unusual phone call from Austin, Texas. The person at the other end of the line claimed to be a friend of one of the survivors of the tragedy. Apparently, he had first called the Austin affiliate of Univision to no avail and then decided to get in touch with the national news desk that broadcasts from South Florida.

One of the survivors, said this "friend," was willing to talk to

Univision and tell the story of what had really gone on inside the trailer, as long as Univision would agree to protect his identity. At the time, none of the other survivors were talking, and this phone call sounded like the kind of opportunity that any journalist would be a fool to turn down, especially at a network that covered immigration as extensively as Univision.

María convinced the phone caller to put his friend, the survivor of the tragedy, on the phone. With some trepidation, a young Central American began speaking into the phone, and after a lengthy conversation with María, agreed to an interview in Austin with the Univision correspondent there, Martín Berlanga. The young man was crying and wanted to talk about the horrible things he had been forced to endure. He swore that there had been a fight inside the trailer, that the 5-year-old boy had been beaten before he died, and that he had almost gotten dismembered himself. He seemed to have a very urgent need to talk about the experience he had been through.

There was one catch, however.

"I don't want my face shown on television," the immigrant told María.

"Well, I can't do that," the Univision producer immediately replied. "You're making some very serious allegations, and if you want to do that you have to be willing to show your face." Finally, the man agreed.

Martín Berlanga, the Univision correspondent in San Antonio at the time, was in Austin that night thanks to a lucky coincidence. He had gone to the state capital to cover a dispute between the Democrat and Republican legislators regarding Texas's electoral districts. María called him on his mobile phone and gave him an address to go to. Martín understood and respected the conditions under which the interview was to be conducted, as per the caller's request. He would not identify the young man by name, but he would be allowed to broadcast his face on TV. What the man told Martín, however, was difficult to believe.

This young Central American man, about 22 years old, with a round, dark face and bright red shirt, told Martín Berlanga a truly chilling story.

"We couldn't take it any more because we didn't have any air to breathe," said the young man on the Univision news program that was broadcast the following day, Friday, May 16, 2003. "Slowly everyone began to lose control. I was unbelievably thirsty. I urinated into my hands and drank my own urine. And then I urinated again and gave my urine to the guy and the two girls nearby [next to me]."

The controversial part came next.

"I heard some guys say that if we wanted to get someone to stop [on the road] we were going to have to do something," he continued, in an almost-neutral Spanish accent, his face distorted by the horrific memory. "And when the kid passed out, they said, 'We're going to take the boy out, even if we have to do it in little pieces, we're going to do it.' And the father said, 'No, no, please don't take my son out, please, not my son . . .' But they wouldn't listen to him. They started hitting him [the boy] against the back of the truck container. They hit him on the head. They hit him, like, two times. They wanted to push him out through [the holes from the tail lights]. [The father] was running out of energy; he had just enough strength to speak. And to cry for his son."

During the interview, the young man insisted that he had

tried to stop the other passengers from hitting the little boy, but that they pushed him aside and threatened him.

"Just when they were about ready to kill me, two girls pushed their way between us, and the men grabbed them, leaving me on the floor. They pounced on those two girls, hit them, hit them and kept on hitting them. They just wouldn't stop.

"This immigrant, without his shirt at this point, was sitting close to the holes in the back of the truck container, where a bit of air was wafting in. He knew that if he moved away, he might die.

"Nobody's getting me away from here," he said to the two who, apparently, wanted to push him out of the way so that they could get closer to the source of fresh air. The shoves, blows, and cries resumed.

"I grabbed that thing, there wasn't anything that could get me to take my hand off that hole."

According to his version of the facts, before the two then managed to dismember the little boy and inflict more damage on him and the girls, the doors were opened. The young Central American man told Berlanga that he went into the gas station convenience store and was the first person to ask for help.

"I need someone to help me—someone's trying to kill me," he told the manager of the store. Just like many of the other immigrants in that truck, he was taken in an ambulance to the Citizens Medical Center in Victoria, Texas.

After an examination revealed nothing more than light wounds on his arms, the result of a purported scuffle inside the truck container, a nurse supposedly told him,

"You ought to get out of here now, while you've got the

chance." The young man apparently called a few people he knew in Austin and asked them to pick him up there. After that, he said, he ran away from the hospital, so as to avoid getting caught by the immigration authorities.

In his news report, Martín Berlanga showed an image of one of the dead bodies, covered by a sheet with two large bloodstains at head level. Was it true? Had a bloody fight broken out inside the trailer? Had someone dared to lay a hand on the child after he passed out but before he died? The blood on the white sheet seemed to corroborate the survivor's tale.

As time went by, however, his version of the story began to seem less and less credible. First of all, it didn't jibe with what the other passengers had told the Mexican consul general in Houston, Eduardo Ibarrola.

"That's not true," the consul general told Berlanga, in an irritated tone of voice. The story also didn't quite fit with the results of the autopsies performed several days later by the forensics team in Austin, Texas, where the bodies were transferred. According to the forensic expert, the cause of death had been asphyxiation and dehydration, not blows to the head or any other part of the body.

But then what was the explanation for the blood on the sheet? Berlanga wanted an answer. Reporters don't like it when people question the information in their news reports. Finally, he got his answer. One of the forensic doctors in Austin explained to him that when a person reaches the level of dehydration that the truck passengers reached, the lungs begin to fill with water, the blood vessels explode, and the mouth and nose begin to excrete a bloody liquid. That might well have explained the blood on the sheets. But then what about the story of the two men who supposedly wanted to rip the little boy apart to get him through the holes in the back of the truck? None of the other survivors mentioned anything remotely like it.

A few days after the report aired, Berlanga telephoned his interviewee, and confronted him with the information the Mexican consul general had given him, including the autopsy results.

"I swear it's true," the young Central American man insisted. But that was the last time they ever spoke on the phone.

"After we questioned him and told him the other version of the story, he never spoke to us again," Berlanga recalls.

After speaking with various survivors, Berlanga is convinced that if some kind of argument or struggle did, in fact, break out inside the trailer, it would have been a fight to get near the holes where the air was coming in, because there wasn't enough room for everyone. There were at least seventy-three people inside that trailer, and the spot near the air holes could accommodate no more than two or three people at a time. But the story about the little boy who had passed out, and the people who supposedly wanted to dismember him—where in the world had that come from?

"How could someone offer such a detailed description of something that happened in total darkness?" Berlanga thought, questioning the authenticity of his interviewee's story.

"But at the same time," he said, "who would ever invent something like that? And why? Maybe he had had a hallucination; that was the conclusion I eventually came to because there just wasn't any proof. Neither the forensics experts in Austin nor the Mexican consul in Houston agreed with the interviewee's version of the story. But it is possible that he had been hallucinating."

Whether he was hallucinating or not, Maria Henao, the Univision producer, thinks that the Central American man turned himself in to the authorities shortly thereafter and received the same immigration benefits as the other survivors did, benefits he would receive until the trial was over.

Alejandro Hernández couldn't take it any more. He had to tell someone—a police officer, an immigration agent—what was happening to him. Alejandro, a family friend of Honduran immigrant María Elena Castro Reyes, had just received a telephone call from Víctor Rodríguez, who told him that he and his wife Ema had Alexis, María Elena's son, with them. Don Víctor and Doña Ema were ready to hand over the child, but only after they received a payment, somewhere between $1,300 and $1,500. The exact amount was still to be determined.

Alejandro had made the payment for the first part of María Elena and Alexis's trip from Brownsville to Houston. María Elena was to make the journey in the tractor-trailer, with the other immigrants, while Alexis would travel with Víctor and Ema, in their minivan. Alejandro already knew that several immigrants had died in the trailer, but that all the women had survived. That had to mean María Elena was alive. Right now, however, his primary concern was Alexis.

At 8:05 a.m. on Thursday, May 15, Alejandro picked up the phone and telephoned the immigration service. A border patrol agent, Arnulfo Cortes, took the call. Without going into formalities, Alejandro explained what was going on: Víctor Rodríguez had called him up and told him that he was willing to hand over the child for money.

Agent Cortes was all too familiar with the Rodríguez family. The previous day, at four in the afternoon, he and Agent Dagoberto Vega had been watching over the Rodríguez house, and had seen Doña Ema leave in a green SUV. According to María Elena Reyes Castro, it was the same Ford Explorer, Texas license plate number J57RRW, that the Rodríguezes had taken to drive her to the spot where she boarded the truck.

After speaking with Alejandro Hernández, Agent Cortes called Don Víctor from a cell phone to make the arrangements to get the child back. Naturally, Cortes did not identify himself as an immigration agent. He told Don Víctor that he was simply a friend of the boy's family. But Don Víctor didn't take the bait.

In any event, they agreed to meet that same day at the Plaza Mall in McAllen, Texas, and Agent Cortes assured Don Víctor that he would pay him the required amount of money in exchange for the child. The Plaza Mall is very big, so they decided to meet in the parking lot in front of the entrance to Dillard's. Agent Cortes made sure to have the place surrounded before Don Víctor got there.

But something didn't sit right with Víctor Rodríguez. He sniffed a trap. There was something he didn't like about how the arrangements had been made—it all seemed too easy. With all the media attention swirling around the case of the immigrants who had died in Victoria, he was sure that the immigration agents were more suspicious than ever. At 55, Don Víctor couldn't take a chance—he'd be risking everything he had. No. No way would he hand over the little boy that day, no way. Instead, he placed a call to Juan Cisneros, who was also identified by the authorities as Juan Carlos Don Juan.

Juan, who had entered the United States in November of 1998, had met Don Víctor at a bar in Brownsville a year and a half after arriving in the country. At the time, he had no idea that such a man, practically a senior citizen, was involved in the business of smuggling undocumented immigrants across the border. Don Víctor never told him but Juan eventually figured it out for himself.

According to Juan, Don Víctor called him and said he would pay him $100 to hand the little boy over to his relatives. It wasn't much money, but apparently Juan needed the cash.

Juan would have to pick up the child at the Rodríguez house in Brownsville and then hand him over at a mall in McAllen. Juan accepted the deal. But first he had to pick up his girlfriend, Erica Cárdenas, and her 5-month-old child. Erica, who was 23 years old at the time, had lived in the Brownsville area all her life and was a U.S. citizen.

After picking Erica and the baby up in a white 1997 Cavalier, Juan drove over to the Rodríguez home. He got out alone, went inside the house, and greeted Don Víctor, who proceeded to tell him exactly what he would have to do. He also warned Juan to be especially careful because he suspected a trap might have been laid. Juan listened to everything Víctor said, but he nevertheless agreed to take the child to McAllen. One hundred dollars wasn't a bad deal. Plus, he wanted to stay on don Víctor's good side—if this job worked out, Víctor might throw a little more work his way in the future, more important work, maybe.

Shortly thereafter, Juan left the house with Alexis and put him in the car. Before leaving, he went back into the Rodríguez house. He had forgotten Alexis's child seat. Once everything was ready, he closed the car door and headed out to Expressway 83, which went straight to McAllen. Juan figured it would take them an hour, an hour and fifteen minutes depending on the traffic. To make sure Alexis wouldn't get hungry en route, Juan and Erica gave him a Gatorade and a burrito, which he ate before falling asleep in the back seat of the car.

They arrived right on time. Juan drove up to the Dillard's entrance, got out of the car and located the man he thought was a friend of little Alexis's family. What he didn't know was that the man was really undercover agent Arnulfo Cortes. The problems started right away. The man didn't have $1,300, the minimum amount he had been told he would have to pay for the boy. Cortes told Juan that he didn't have all the money on him, which Juan couldn't accept, so he picked up his phone and dialed Víctor's number. After explaining the situation, Don Víctor spoke directly with Cortes. No, Cortes said, he couldn't pay the whole amount. Just $300 for the moment and the rest a little later on that same day. Don Víctor accepted the new offer and told Juan to go ahead with the exchange.

Juan took the money and went over to the car where Erica was sitting with her baby. Then he walked Alexis out and handed him over to the person he had assumed was a relative or family friend. Once the job was done, Juan got in his car and tried to get out of the mall as quickly as possible. That was when he realized that he had walked into a trap, just as Don Víctor had warned him. All the exit lanes at the mall were blocked off by McAllen police cars as well as several immigration service vehicles. In a moment of desperation, Juan got out of the car and began to run, leaving Erica and the baby alone in the car. But he didn't get very far. He was stopped and arrested on the spot. Neither Juan nor Erica were ever linked to the overall scheme that ended in tragedy two days later.

That afternoon, seeing that Juan had not made it back to Brownsville nor had he telephoned, Doña Ema and Don Víctor launched their escape plan. Just as Don Víctor had suspected, a trap had been laid for him. Now they were going to have to get out of Brownsville as quickly as possible. Without a doubt, the best escape route they had was Mexico.

At least two telephone calls might have averted the colossal dimensions of this tragedy. One of them had been placed from inside the trailer by a young Honduran man with a mobile phone. Shortly after 11 p.m. on the night of May 13 he had dialed 911. But Matías didn't speak English, and the operator did not speak Spanish, so the call had not been reported to any of the South Texas police departments.

The other phone call was placed just before midnight by a man who had noticed something amiss on a trailer with New York plates that was traveling through the city of Kingsville. In what was a non-emergency phone call, he told the operator that he had seen someone stick a hand out of a hole in the back of a trailer, waving a handkerchief. The operator did not contact the police department at Kingsville nor did she call the one at Bishop, to the north of Kingsville. On Monday, June 30, 2003, the Kingsville city authorities announced that the operator who had taken the non-emergency phone call had been suspended with pay.

If one of the operators who had received the call from Matías or the unidentified driver had alerted someone, everything would have turned out differently.

"Carolina Zaragoza [the Mexican consul in charge of civil protection] was our guardian angel," remembers Alberto, one of the immigrants from Mexico. "We watched her working. She always told us the truth. She would say, 'OK, this is going to take a while. I promise you that we are going to stand by you. We are not going to let you go through this alone. We have a lot of people working on this.' They [Mexican consuls Carolina Zaragoza and Eduardo Ibarrola] fought hard to keep us from ending up in some immigration service detention center. She would talk with my family and then tell us what they told her."

Thanks to the intervention of the Mexican consulate in Houston, the immigrants were able to avoid getting deported until after the trial of the people accused of orchestrating their ill-fated trip.

"The U.S. immigration authorities handled the situation in a really humane way," Consul Zaragoza remembers. "The negotiation was basically about granting the survivors 'migrant' status so that they could work while the judicial process ran its course. The U.S. needed them as witnesses for the prosecution."

The survivors, however, were far from the Mexican consulate's only problem. They also had to take care of the dead.

They had to be certain, beyond all doubt, of the identities of the people who had perished. They couldn't possibly call a family without being completely certain that the body had been positively identified. Two of the easiest bodies to identify were those of Marco Antonio and his father. The father had brought both their birth certificates with him. The real problem was identifying the immigrants who had died without any identification on them. Many of them, out of desperation from the suffocating heat, had placed all their belongings on the floor of the trailer.

The bodies of the nineteen mortal victims were taken to the forensic services at Austin, the state capital, and after being examined they were sent to the Compean funeral home in Houston. As all this was happening, a team from the Mexican consulate in Houston began sorting through the deluge of photographs they had received (not to mention faxes and telephone descriptions) from the families that feared their relatives had died, in an effort to identify the bodies of the dead. The process was going to be long and painful.

"Ma'am, listen, you know what happened, they were in the trailer . . ." So began the conversation between Carolina Zarazoga, the Mexican consul, and another Carolina, Marco Antonio's mother. Zaragoza had been preparing her, leading up to this, to eventually tell her the inevitable. "In the end, unfortunately, Marco didn't make it."

"Marco didn't make it." With that sentence, Carolina's worst fears were confirmed. For several days, she had held onto the hope that Marco Antonio hadn't been one of the victims.

"It can't be true; it can't be my son," gasped Carolina, crying over the phone to the Mexican consul. Zaragoza, momentarily abandoning her role as a Mexican government employee, tried to comfort Carolina, mother to mother:

"There is nothing worse than losing a child . . ." But nothing could assuage Carolina's anguish. Nothing. In reality, Carolina was still trying to believe that this was just a huge mistake and that her son Marco Antonio would turn up sooner or later.

"I send you my deepest condolences," Consul Zaragoza said. "Marco Antonio is leaving here right now. He is taking off and will arrive in Mexico at 3:15 in the afternoon." Thanks to her wonderful manner, her professionalism, and the accurate information she always gave, Carolina Zaragoza was one of the few people that Carolina truly trusted.

Marco Antonio's body arrived at Mexico City in an Air Force plane along with the bodies of ten other Mexican immigrants who had died in Victoria. Of the remaining five, one was sent by car to Linares, Nuevo León, and the other was buried close to Houston. The other three bodies would be sent to Mexico later on. The Mexican government paid for all these expenses.

"I really just wanted to see him," Carolina remembers. "I wanted to make sure that it was really my son." But Carolina didn't go to the airport. As soon as the Air Force plane landed at Benito Juárez airport in Mexico City, the eleven bodies were placed in hearses and brought to a funeral parlor in Colonia Roma, a neighborhood in the city's center.

When Carolina arrived at her son's wake, someone led her over to the tiny closed coffin where Marco Antonio lay. Even then, she still clung to the idea that maybe it was all a mistake, an error—anything at all so that she might be able to hang onto the hope of seeing her child alive.

"I want to see Marco," Carolina said to her brother Natalio, who went over to the coffin with her. Together, they opened it.

"I opened the lid," Carolina recalls, "and oh, I see a big child with make-up on. Marco was very white, so pretty. But [the body] seemed darker than I was. I said to my brother, 'No, this isn't my son, this isn't my son.' But it was. I touched his fingers. I ran my hand over the two beauty marks he had. His little feet were covered up, but I touched them with my fingers, and I gave them a little massage. They were my two little mangos. They were beautiful. His hands, too, were so pretty. This was my son! This was Marco Antonio! I wanted to shout and cry and beg someone to give my son his life back! This was my only son! Nobody answered me, but it was Marco. And I said to God, 'God, God, I know you are with me, make me strong, God, I know you have mercy on us, hallowed be thy holy name, God, and if I ever did anything wrong, God, please forgive me and receive my son.' And then I sang Marco a little song, one that he had made up. He used to sing it to me: 'Oh, your baby loves you, oh, your Mommy loves you, too . . .' And I sat down and talked to him. 'Oh, my darling,' I told him, 'you're home now.' Very gently I touched his hands, because he said to me, 'Make nice to me, Mommy, and let your hair grow out so that you can wear it in a ponytail.' And I said to him, 'It's all right, my darling, you are beautiful and you will always be in my heart.'"

7

KARLA AND
THE TRIAL

Karla Chavez spent the early morning hours of May 14 at a hotel in Victoria, Texas, where she must have seen the story of the tragedy on TV. After watching the news, she was probably scared that she would be accused of being responsible for their deaths. Very quickly, she came up with a plan to flee the United States and head for Honduras.

First, Karla went back to the Harlingen suburb of Combes, to the house that she and her boyfriend, "El Morro," had bought for around $50,000. According to the Associated Press, she rapidly packed four bags, grabbed her three children, and told her neighbors that she had to go to Honduras because her parents were ill. Karla left behind several vans, her children's bicycles, and her children's dog, which she chained up alongside a bag of Pedigree dog food. The dog had recently given birth to puppies, but nobody knows what happened to them. That was the rather abrupt manner in which Karla ended her eight-year stay in the United States.

Long before, when she was a teenager, Karla had crossed the U.S. border illegally and had gone to work at a Levi's factory. She eventually settled down in the U.S. and had three children, but she now found herself forced to give it all up and flee back to her home country.

But Karla didn't feel very safe even after she returned to Honduras. The U.S. Spanish-language TV stations broadcast their news programs in Honduras, and she knew that the case had quickly become one of the most talked-about immigrant tragedies ever to occur on American soil. No. She was convinced that they wouldn't leave her alone in Honduras. Since she had left, the death toll had risen to nineteen, and anyway, it would be easy enough to trace her mother's address from her U.S. telephone bills.

According to *The New York Times,* Karla spent twenty-three days holed up inside the home of her brother Carlos Alberto in the city of San Pedro Sula. Then, she left for Guatemala, placing her three children in the care of her sister-in-law and her mother. But they were already waiting for her in Guatemala.

The international investigator at ICE, Byron Lemus, never lost track of Karla. The ICE is the new immigration service that now operates under the authority of Homeland Security.

One month after the immigrant deaths in Victoria, Texas, Karla was detained as she tried to cross the Honduras-Guatemala border. Investigator Lemus's work had been effective. Very effective. The diplomats had done their job, as well, convincing the Honduran government to deport Karla to Honduras via Houston, Texas. This was a legal maneuver that allowed the prosecutors to charge Karla with the crimes she purportedly committed while she was on U.S. territory. Escorted by Investigator Lemus, the ICE supervising agent in

Guatemala City, and a representative of the Guatemalan government, Karla traveled by plane from Guatemala City to Houston, Texas.

There was no escape for Karla.

As soon as she arrived in Houston on Saturday, June 14, 2003, the one-month anniversary of the tragedy, Karla was presented with a warrant for her arrest and was taken to the Justice Department offices in the same city. Upon arrival, she was photographed, fingerprinted, and brought up to the second floor of the building for interrogation. There, two people were waiting for her: agents Marc Sanders and Deanna McCarthy. Karla told them that she would prefer speaking in Spanish because her English wasn't very good. But she was very thirsty and requested a bottle of water.

Karla had been living in fear of this moment for an entire month. And it had finally arrived.

At five in the afternoon that day, the conversation began. First, the agents immediately told Karla that an investigation was underway. They also informed her that several people had made statements that identified her as the person responsible for what had happened on May 13 and 14 in Victoria, Texas. Karla didn't say a word. The agents wanted her testimony, but they also had to let her know her rights. After listening to what they had to say, Karla signed the documents they placed in front of her and, according to the investigation report from that same day, she voluntarily gave up her rights and agreed to talk to them without having an attorney present. Nobody from the Honduran consulate in Houston was present, either.

Karla didn't waste time. The first thing she told them was

that she was not responsible for the tragedy in Victoria, and she went on to say that other people had made the arrangements to transport the immigrants. But they didn't believe her. Agent Sanders told her that several immigrants and some other suspects had confirmed that she had been the focal point of the operation.

Karla, of course, had no idea who had spoken out against her. Nevertheless, she did tell the agents that Abel Flores had made all the arrangements to transport the immigrants from Harlingen to Houston, and that Flores and the other people involved in the operation were merely friends of hers. Once again, the agents were not convinced by Karla's arguments.

"If you really are innocent," they asked her, "then why did you leave the country?"

"Because I was afraid I would be arrested for being friends with some of the people involved in the case," she replied.

"You're lying," said Sanders.

When Karla was arrested in Guatemala, she was found carrying a notebook with several telephone numbers, plus a receipt for a $3,000 payment she had made to the lawyers representing the father of her children, Arturo Maldonado (also known as Heriberto Flores), who was currently doing time in a Texas prison. Karla had no idea what other evidence they had against her, but at that moment she was more concerned about calling her mother to explain where she was. So far, she hadn't told anyone that she had been arrested, and she asked permission to make a long-distance call to her mother's home in Honduras. The agents warned her that the phone call would be recorded, and they asked for her consent. The report, once again, indicates that she agreed to this.

While the agents prepared the audio equipment to tape Karla's phone call to Honduras, Karla was shown a series of photographs and was asked to identify all the people she knew. Karla did not recognize Tyrone Williams or his companion Fatima Holloway, but the minute she saw the photographs of Doña Ema, she began to cry. Even though she had already named Abel Flores, she really didn't want to identify anyone. But the agents pressured her and insisted that they needed her cooperation to see if she was really telling the truth about what had happened in Victoria, Texas. After they gave her some Kleenex, Karla pointed to photograph number 5 of the six snapshots laid out before her. It was of Ema Rodríguez, Víctor Rodríguez's wife. Ema had been the one who had phoned Karla and informed her that a number of immigrants had died in the trailer, and that eleven of them were their clients.

The agents were well-prepared; they had been planning this interrogation for almost a month. Next, they showed her another series of photographs, among which Karla identified her friend who owned the Lincoln Navigator Karla had been driving and whose name was on the various wire transfers that several immigrants had made as payment for their trip. The agents had found all this information with little or no problem at all.

Karla, however, told the agents that her friend hadn't been involved at all in the immigrant trafficking operation. The telephone was still not ready at this point, and the pressure was mounting, for they had been questioning Karla for almost two hours by now. Karla asked to go to the bathroom again; when she emerged, she was told that she could finally place her phone call to Honduras. It was around seven, Saturday night. Someone had to be at home. Karla dialed a number in Honduras, but nobody answered, just a voice on an answering machine. After a while, Karla finally got through to her mother and explained that she was not in Guatemala but in Houston, where she had been arrested. Look after the children, she told her mother, but don't tell them that they caught me. Her mother, shocked by the fact that her daughter was in the United States and not in neighboring Guatemala, told Karla that she was coming to Houston to find her a defense lawyer. The phone call lasted less than twenty-five minutes.

After Karla hung up, the interrogation continued. Her day wasn't over yet.

Next, Agent Steven Greenwell arrived to continue the questioning session. Agent Deanna McCarthy, the other woman in the group, introduced the two and then left. With Greenwell, Karla began to identify the photographs of the people she knew, one by one.

The first person she identified was Víctor Rodríguez, Doña Ema's husband. Then there was Abel Flores, the man in charge of transporting the immigrants from the safe houses to the truck. According to Karla, Víctor was the person who had closed the trailer doors. A bit later on, she would also identify the photograph of Alfredo "Freddy" Giovanni García, an associate of Abel's with whom she had been romantically involved.

Karla also identified Gabriel Gómez, the third in a series of six photographs who she said had sent fourteen immigrants to the tractor-trailer. Gómez had also fled the United States and, in fact, had been with Karla when she was arrested in Guatemala. "Tavo" Torres, photograph number 4, had sent twenty immigrants to the truck, according to Karla.

Karla must have been exhausted at this point, but the agents wanted to keep going with the interrogation. It was the first time they had ever spoken to her, and without a lawyer present they could ask her whatever they wanted. Their supervisor, Gus Meza, came by to make sure the conversation was still productive, and he asked Karla a few questions about her mobile phones. Karla then confirmed that she had two mobile phone numbers, one for Abel and one for her. That was how Abel had kept her informed about the truck's progress on the road: after conferring with the driver, he would call Karla and relay the information.

Now, with Supervisor Meza, Karla was told about a certain receipt from a hotel in Victoria, Texas, indicating that she had spent the night of May 14th there. There wasn't much more she could hide at this point. Karla then told the agents that she had been romantically involved with "Freddy" and that when Doña Ema told her about what had happened inside the truck, she and "Freddy" left Corpus Christi, intending to reach Victoria between two and three in the morning on May 14.

As the evening wore on, Karla gave them more names: Rafa "La Canica" had placed seven or eight immigrants in the truck; "El Caballo" had sent three; Ricardo Uresti had sent five, including a child and his father; Salvador "Chavo" Ortega had also sent some undocumented immigrants, though she didn't know how many.

As far as the agents were concerned, it was very clear that this had not been Karla's first immigrant trafficking operation. In fact, she herself admitted that she had participated in at least four other operations transporting more than sixty-two people. This was exactly what Supervisor Gus Meza wanted to hear. The agents asked Karla if she would be willing to recount her version of the facts in front of a video camera. After all, it would only be the same story she had just told them over the past four hours. She accepted, and on Saturday, June 14, 2003, at 9:19 in the evening, Karla gave her testimony in front of a video camera.

Later on, all this testimony would be used against her. And if the Attorney General of the United States, John Ashcroft, gave his approval, Karla could possibly end up facing the death penalty in the event she was found guilty.

By the time Karla finally contracted a lawyer, she was already mired in very serious legal problems. The statements she had made that Saturday, without an attorney present and without a representative from the Honduran consulate in Houston, had all but buried her.

The prosecutors for the case had a substantial amount of evidence against Karla, including statements made by some of the survivors. On Wednesday, May 13, Lupita Gorman, the agent from ICE, the newly-created Bureau of Immigration and Customs Enforcement, paid a visit to the Victoria community center, where she met several survivors of the tragedy. There, Agent Gorman was able to interview three Salvadorans, among others, who clearly identified Karla Chavez as the person who had coordinated their trips from El Salvador to the United States. They also acknowledged having met her in person at a safe house in Harlingen, Texas.

Agent Gorman had brought with her six photographs of women who looked somewhat like Karla Chavez. Each photograph had a number. In the course of her conversation with Salvadoran survivor Ana Márquez Aguiluz, Agent Gorman asked

Ana to identify Karla Chavez from among the six photographs. Ana pointed to photograph number 4, as did two other Salvadorans who had survived the tragedy, Carmen Díaz Márquez and José Martínez Zúñiga.

Lorena Osorio Méndez, a native of Mexico, also spoke with Agent Gorman, and told her how she had contacted Karla to arrange her trip across the border, which was to cost her $1,900. Lorena crossed the border close to the town of Rio Grande City, and then was taken to a house in Harlingen where she had to pay Karla a $1,000 advance; the other $900 would have been due upon arrival in Houston. When Agent Gorman asked Lorena to identify Karla from among the six photographs she had brought with her, Lorena pointed her finger at photograph number 4.

Before going down to the community center in Victoria, Agent Gorman had gone through an archive in search of photographs of women who bore a resemblance to Karla Chavez. Karla, she decided, would be photograph number 4.

"I am innocent," Karla Chavez told the Consul General of Honduras in Houston, Lastenia Pineda.

Fifteen Hondurans had been traveling in the trailer. Fourteen had survived, and one man, 32 years old, had died of asphyxiation. Consul Pineda calculated that each passenger had paid between $2,000 and $5,000 to get from Honduras to the U.S. In general the journey was made entirely over land, through Guatemala, Mexico, and then on toward the southern United States.

"Things have been very difficult in Honduras recently," said

Consul Pineda. "People come here hoping to find work—they all know that here [in the U.S.] they can make money." The consul stated that a Honduran country laborer might earn some 25 *lempiras* a day—about $1.50 according to the 2003 exchange rate. In other words, after ten or fifteen minutes on the job in the United States, Hondurans can earn the same amount that would take them all day to earn in their home country. Many of them arrive with the hope of earning enough money to build a house and eventually return home. But in the wake of the terrorist attacks of September 11, 2001, that circular immigration cycle began to break down.

Oddly, for the fifteen Hondurans traveling in the tractor-trailer, the person who was supposedly helping them get up to Houston was, ironically, a fellow Honduran. Consul Pineda was one of the few people who got a chance to speak with Karla before the trial. When foreigners are charged with crimes in the U.S., the federal government is required by law to grant them access to the consulate of their home country. As such, Karla was allowed to speak to Consul Pineda on at least three occasions. The consul described her as "calm" during her time in jail, "although sometimes she begins to cry; she feels sad; she misses her children; she is a good mother." During those visits, Karla was alone in her cell. "At some moments, she seems very serene, and then at other times you can see that she is suffering terribly inside. You can tell she misses [her children] terribly."

Karla, the mother of three children, had come to the United States when she was 17, and already had her permanent residence card. Eight years after her arrival, at age 25, she was arrested.

"She says she is innocent," Consul Pineda says. "That is what she has always told me. She insists that she is not responsible [for the immigrant trafficking operation]. She does not place the blame on any one specific person. When we talk, she says to me, 'I am innocent; I am not involved; I am not responsible for this.' That is what she has said to me. She gives the impression that she is not conscious of what she is going through."

Karla Chavez, who at one point feared she might face a death sentence, was finally charged with fifty-six counts and the authorities referred to her as the ringleader of a group of coyotes coordinating the journey that ended in the deaths of nineteen immigrants. But Consul Pineda always remained very skeptical of these charges.

"The truth is, she's an awfully young woman to have been the ringleader. She would have had to be a genius to control such a large network of operations. Personally, as consul and as a human being, I think she would have had to have been a genius, and I don't believe it. I don't think she's that competent. I have interviewed her two, three times, and she's just a girl. She is a young woman who—and I am not defending her when I say this—I just don't think had the capacity [to do this]."

For months, the prosecutors toyed with the idea of seeking the death penalty for Karla. After all, she did seem to be the person who had the most connections to all the other defendants in the case. Plus, she was under arrest.

Despite all this, six months after the incident in Victoria, the Attorney General of the United States, John Ashcroft, decided not to seek the death penalty for Karla. But Ashcroft himself did not make the announcement to the press on Monday, December 1, 2003. The task fell instead to Michael Shelby, the federal prosecutor in charge of Texas's southern district, who had taken over the Victoria case.

"Federal law requires the government to prove beyond a reasonable doubt that the defendant intended to cause the death of a victim before a death penalty may be imposed. We have determined that such proof is presently unavailable."

Shelby's announcement was about the only good news Karla had received over the past six months. In addition, said the *Houston Chronicle,* prosecutors would not be seeking the death penalty for four other defendants, either: Claudia Carrizales, accused of using her apartment to hide undocumented immigrants; Abelardo Flores, who had allegedly located the vehicle that would transport the immigrants from Harlingen to Houston; Norma González, who was linked to the plans for transporting two undocumented immigrants (one of whom died); and Víctor Jesús Rodríguez, who authorities say had helped smuggle Latin Americans into the United States along with his parents, Víctor and Ema Rodríguez. The charges against Ms. Carrizales were dropped after the judge determined there was insufficient evidence.

The tragedy in Victoria created a serious public relations problem for the Ministry of the Interior in Mexico. It soon became very clear that all the immigrants who died in Texas were either Mexicans or had at least passed through the country, and it was also quite evident that the authorities had done nothing to stop them. Mexico could not pretend that this was a problem the U.S. government had to handle on its own. The Fox administration had to make some kind of statement to show that Mexico, too, was committed to the fight against immigrant trafficking, but the real story is that Mexican immigrants in the United States generate $14 billion a year in income for the Mexican economy.

The Mexican government has never, under any administration, made any kind of serious effort to stop the flow of undocumented immigrants to the north. Never. And that is because it wasn't—and isn't—in the country's best interest. The main focus of President Vicente Fox's electoral campaign was that of informing the populace about the dangers of crossing the U.S. border illegally, but the end result was nothing more than thousands of posters which were printed up and tacked on electricity poles and bus station walls all along the border.

Without the money sent home by Mexican immigrants living in the United States, Mexico's economy would crumble. But this was not the right moment to talk about economics or immigration agreements. And the well-trained political nose of Santiago Creel Miranda, Mexico's Secretary of the Interior, whom many people looked upon as a possible presidential candidate for the Mexican elections in 2006, sniffed this out immediately.

On Tuesday, August 12, 2003, the smiling face of Secretary Creel, framed by his well-clipped salt-and-pepper beard, was photographed by dozens of reporters who had been summoned to a press conference in Mexico City. In a televised image that was broadcast throughout Mexico and the United States, Creel was shown shaking hands with the Attorney General of Mexico, General Rafael Macedo de la Concha. Their faces beamed with satisfaction; they were there to announce the arrest of twelve immigrant traffickers in a police operation carried out in the states of Guanajuato, Nuevo León, Tamaulipas, and San Luis Potosí.

On top of this, orders were underway to apprehend twenty-five additional people who might have been involved in the Victoria case. Among those arrested were Eliseo and Ismael Peralta. According to Creel and Macedo, the Peralta brothers had placed several immigrants on a bus from Guanajuato to the U.S. border, from which point the immigrants then boarded the trailer.

Creel, repeating the accusations of the U.S. authorities, assured journalists that Karla Chavez was the region's pre-eminent immigrant trafficker, and added that the Mexican government was also investigating the question of whether or not she had been involved in the deaths of the eleven Mexican and Central American immigrants who had been found inside a railway car in Denison, Iowa, in October of 2002.

The message Creel wanted to send out—specifically to the United States—was that Mexico was pulling its weight in the fight against immigrant trafficking. And he had several arrests to prove it. Later on, of course, they would proudly point to the investigation and arrest of two suspected kingpins: Víctor and Ema Rodríguez.

The Rodríguez family never really acted much like a family. When they sensed that the authorities were on their trail, following the arrest of Juan Cisneros and Erica Cárdenas at a

McAllen mall, Víctor and Ema fled to Mexico. The fact that Juan and Erica had been arrested just after Juan handed over the 3-year-old child of one of the Honduran immigrants who had been traveling in the trailer was a very clear sign that they were under surveillance. But they left for Mexico without their son, Víctor Jesús, who was apprehended in the Rio Grande Valley shortly after the incident in Victoria.

Víctor Jesús was one of the nine people arrested in the United States for suspected involvement in the Victoria tragedy. But five other people were still at large, including Víctor Jesús's parents. In some way or another, all fourteen people the U.S. government charged in relation to this case were connected to Karla Chavez and a sheaf of Western Union receipts that were evidence of payments for the transfer of undocumented immigrants.

Víctor and Ema Rodríguez were not very imaginative. Or perhaps they assumed that the Mexican authorities were corrupt and incompetent, and that none of the people arrested in the U.S. would speak out against them or give the authorities information that might lead to them. But they were very wrong. Karla, for one, had spoken with the investigators, and the immigrants were talking, as well.

It didn't take too long to find out that several of the immigrants traveling in the trailer had been rounded up at a house in Matehuala, in the Mexican state of San Luis Potosí, before being sent on to Matamoros, near the U.S. border. The Mexican police, working on information provided by the U.S. authorities, started their investigation right there, in those two cities. Easy as anything.

On September 5, 2003, Ema Rodríguez was arrested in

Matehuala, in the company of Rosa Sarrata, whom U.S. prosecutors charged with harboring undocumented immigrants at a house in San Benito, Texas. Víctor Rodríguez, however, was not arrested. Once again, the Rodríguez family was not acting much like a family, and because of this, it was that much more difficult to arrest him.

The Mexican investigation team now turned its attention to Matamoros. Their reasoning: Víctor Rodríguez was an older man, and they felt it extremely unlikely that he would suddenly start changing his habits or going to places other than the usual spots he liked to frequent. Even though he was a U.S. citizen, Víctor Rodríguez still traveled around the border zone like any other Mexican. And that was precisely how he was caught.

During Easter Week of 2004, a Mexican newspaper reported that Víctor Rodríguez had been arrested in Mexico. But since almost nobody works during Easter Week in Mexico, the newspaper was unable to find out any of the details regarding his capture. Not until the Monday after Easter did the news break: Víctor Rodríguez had in fact been arrested several weeks earlier, on March 29th, in Matamoros, Tamaulipas. Following Víctor Rodríguez's arrest, Karla's lawyer, John LaGrappe, made the following statement to the *Houston Chronicle,* a newspaper that closely followed the Victoria case and offered exhaustive coverage of the entire process:

"I am absolutely convinced Víctor Rodríguez was the real ringleader," LaGrappe said.

The arrest, however, raised several questions. Why hadn't the Mexican authorities announced it to the press earlier? And if the U.S. prosecutors had been duly informed, why hadn't *they* announced the arrest of such a key figure in the case?

These questions reflect the very serious conflict between the U.S. and Mexico regarding the investigation of the case and the procedure that was followed. Mexican law does not permit the extradition to the United States of any person who might be condemned to death or life imprisonment. The maximum prison sentence in Mexico for this type of crime is twenty-eight years. Above and beyond those limitations, the Mexican government was not willing to extradite these traffickers to the U.S. First they would have to be tried in Mexico. Only after that would the Mexican authorities decide what to do with them.

For the U.S. prosecutors, the arrests made by the Mexican police seriously complicated the situation of the other detainees in the United States. It didn't seem fair to them that someone arrested in Mexico would face between ten and twenty-eight years in prison, while those arrested in the U.S. could be put behind bars for the rest of their natural lives. And yet no formal attempts were ever made to seek the extradition of those arrested in Mexico. Despite this very obviously frustrating situation, the U.S. prosecutors allowed the Mexican detainees to be tried first under Mexican law.

With the arrest of Víctor Rodríguez, the Mexican government now had four of the fourteen people that the United States sought in relation to this case. In addition to the arrests of Ema Rodríguez and Rosa Sarrata—accused of using a safe house in San Benito, Texas, to hide undocumented immigrants—the Mexican government now had custody of Octavio "Tavo" Torres, identified by Karla Chavez as one of the coyotes who sent a number of undocumented immigrants to the trailer.

Despite the tensions between the two countries, the Mexican and U.S. governments did agree on something: they both wanted to blame the tragedy on the coyotes, the traffickers of undocumented immigrants. It was never even suggested that the Mexican and U.S. governments might bear any responsibility in this case—for example, as a result of the dismal economic situation in Mexico or the lackluster immigration policy enforced by the United States. No. As far as they were concerned, the coyotes were entirely to blame.

"What the Mexican government has always said, and on a personal level I agree with them, is that the traffickers have no respect whatsoever for the life or the well-being of the human beings with whom they operate their trafficking ring," said the Consul General of Mexico in Houston, Eduardo Ibarrola. "Many people would like to see these traffickers as heroes who help the jobless find work in other countries, but that is simply not the case. The truth is that they transport them in sub-human conditions. And what I saw, the bodies lying in that trailer in Victoria, is absolutely unacceptable and completely deplorable. The trafficking of human beings is a punishable crime in the United States, and it is a federal crime in Mexico. As authorities, we must fulfill our obligation and enforce the law."

From the beginning, the Mexican and U.S. governments made Karla out to be the devil incarnate. It is possible that Karla and several of the other defendants caused the death of those nineteen immigrants. It is possible, but it was never their intention. The coyotes, or *polleros,* were being exclusively blamed for the deaths of the undocumented immigrants, and that was what the defendants and their lawyers found most unjust, because in reality the responsibility was a shared one. The governments of Mexico and the United States were also partially to blame for what happened.

Coyotes had long since become a necessity for anyone who wanted to cross the border illegally into the U.S. Because of the ever-increasing surveillance along the border—especially after the terrorist attacks that took place on September 11, 2001—it became very difficult for people to enter the U.S. on their own. That is the reason why immigrants were—and are—so willing to pay coyotes thousands of dollars per person. They need someone to help them across the border. The problem, however, is that over time methods and routes across the border have grown increasingly dangerous.

In the past, people could and often did enter the United States near the border towns, but now they were finding themselves forced to travel through white-hot deserts, rugged mountains, and a rapidly rushing river. Once inside the U.S., they would often have to travel in sealed truck trailers, railway cars that locked from the outside, and via other exceedingly unsound transportation methods just to get away from the border. That was why so many immigrants were dying.

The coyote business had blossomed as the result of the U.S.'s very flawed immigration policies, Mexico's permanent state of economic crisis, and both countries' inability to reach any kind of immigration agreement. If, instead of hunting down immigrants and penalizing illegal border crossings, both governments could find a way to regularize the entry of immigrants in an orderly fashion so that Mexico might provide the U.S. economy with the workers it needs, border deaths would become a thing of the past, and the countries would finally legalize something that occurs every single day, regardless of the law.

U.S. prosecutors could not seek the death penalty for Karla Chavez. They could, however, in the case of driver Tyrone Williams, because he had fled the spot where some of the immigrants had died and others very nearly expired as well. According to the prosecutors, Williams was guilty of transporting undocumented immigrants and leaving them in the back of a tractor-trailer near a gas station in Victoria, Texas, without seeking help from the authorities.

Why did he flee? He panicked. His actions were illogical: how could he have possibly thought they wouldn't track him down if the truck and the license plates to the trailer were registered in his name? He fled, simply, because he didn't know what else to do. Then, in a wave of remorse, he told his story to a nurse in a Houston hospital. By then, of course, it was too late. Shortly thereafter he was arrested in the hospital where he was being treated.

"When an act directly results in the single largest loss of life in any contemporary smuggling operation, justice and the law demand the accused face the ultimate punishment," said Michael Shelby on March 16, 2004, justifying the Justice Department's decision to seek the death penalty for Tyrone Williams. Craig Washington, Williams's lawyer, was naturally opposed to the sentence his client was going to have to face.

"I truly believe that those who are on the jury will believe it would be wrong to give Mr. Williams the death sentence," said Washington.

Though not a formal charge, there was another suspicion that Williams would have to contend with: that perhaps he had heard the immigrants' cries and constant banging against the walls of the truck container and had not responded. It would be impossible to prove one way or the other. Only Williams and his companion, Fatima Holloway, know what they heard.

Williams's problem was that on the day of his arrest, he admitted to the police that he had heard the immigrants board the truck trailer. Now, if he had heard them climb into the trailer, how was it possible that he hadn't heard them during the ride? It is possible that the engine and highway noise blocked out the sounds coming from inside the trailer. But several immigrants testified that they made a tremendous amount of noise inside that trailer, slamming themselves against the wall and moving from one side to the other in an effort to get the driver's attention.

Did Williams realize that his passengers were trying to communicate with him? Is it true that he didn't know his tail lights were broken until he reached Victoria? There is no way to prove that Williams had any criminal intent, but the U.S. Attorney may try to prove that he had been negligent on the basis of the fact that he did not respond to the many cries for help that came from the inside of the trailer and also that he fled the scene when he saw how many people had died.

If the immigrants made a lot of noise to get Williams's attention, it was only after passing the patrol booth at Sarita, Texas. That, at least, is the opinion of border patrol agent Xavier Ríos.

"Believe me," he told the *San Antonio Express-News*. "If they were making noise when it [the truck] came through the checkpoint, it would've been pulled and searched." This tends to corroborate the version offered by the majority of the immigrants, who said they decided not to make any noise when they passed the checkpoint at Sarita, despite the fact that many of them were already exhibiting very clear symptoms of asphyxiation and de-hydration.

Karla Chavez stated that she had been working for Abel Flores. And Abel stated that he had been working for Karla. The evidence shows that they had been working together. Because of the evidence against both of them—the fact that they had fled, the immigrants' testimony, the Western Union receipts, the statements made by driver Tyrone Williams—neither of them thought that a jury would declare them innocent. Karla had already spoken with the authorities on at least one occasion about Abel's role in the operation. Now it was Abel's turn.

About two months before the trial, scheduled for the summer of 2004, the prosecutors for the case announced that Abel Flores had agreed to plead guilty of harboring and transporting undocumented immigrants who died and/or suffered physical harm. By pleading guilty, Abel would face a maximum sentence of life in prison and a fine of $250,000. Why did he do it? What did he stand to gain? The person who had the answer to that was Assistant U.S. Attorney Daniel Rodríguez, who told the press that if Abel cooperated in the trial against the other defendants, including Karla, he would ask the judge to give him a reduced sentence.

This was an important legal victory for the prosecution; not only had they elicited a guilty plea from one of the defendants, but in the process they broke the code of secrecy and silence that had previously prevailed among the group of traffickers. Now that Abel was willing to testify against his former colleagues, the rest of the defendants would soon start pointing fingers in every direction imaginable in order to save their own skin. In the end, the prosecution's strategy—to create an atmosphere of mistrust among the defendants—would result in prison sentences for several of the accused.

The defense attorneys, of course, were aware of all this and from the beginning had tried to avoid a group trial. With individual trials they would have a better shot at defending their clients and might be able to offer them a bit of hope for freedom after serving a long sentence.

Just like Abel, Fatima Holloway, Williams's driving companion, pled guilty in exchange for a reduced sentence.

Following the arrest of Norma Gonzalez in May of 2003, prosecutors discovered that she had an extensive criminal record. After she was taken into custody, Norma denied being involved in an immigrant trafficking operation and instead told the agents who apprehended her that she was just a "referral" and that she had been involved in the trafficking of undocumented immigrants for less than a year.

According to court documents, Norma acknowledged that she had accepted a payment for helping someone across the border, the brother of a woman who worked at the restaurant she managed. But she also said that she had simply passed the payment along to the real traffickers.

Very quickly, however, they caught Norma contradicting herself. Norma's own husband told the ICE agents that she earned $100 for every immigrant she brought in, and that she had sent two undocumented immigrants to the tractor-trailer that ended up in Victoria. A Western Union receipt for the payment of one of those two immigrants was found on the floor of Norma's husband's car. Norma's husband confided to the agents that he had wanted his wife to get out of the business but that she never paid any attention to his concerns.

This was not the first time Norma had been charged with involvement in criminal activities. In 1985 she had been accused of robbery; in 1998 she had been linked to coyote Gelacio Hernández's immigrant trafficking network and was charged with receiving payment for six undocumented immigrants who had wanted to travel from Texas to North Carolina, although the case against her was later dropped; and in 1999 and 2002 she was accused of assault with intent to harm.

The most serious item in Norma's police record was the charge that she had lied to obtain her U.S. citizenship. The documents used in court indicate that Norma was arrested by border police in March of 1990. Under an assumed name, Norma had claimed to be a native of El Salvador and was subsequently deported. But her photograph and fingerprints had been put on record. The photograph and the fingerprints that had been taken then were identical to those of the woman arrested by ICE in 2003.

When Norma applied for her green card, she stated under oath that she had never been arrested. Her record proved that this was not true. Years later, Norma became a U.S. citizen.

And now, for not having told the full truth, ICE began the process to revoke her U.S. citizenship. In the end, this detail would prove to be slightly less important to Norma, given that she now faced the possibility of spending the rest of her life in jail.

On May 15, 2003, Erica Cárdenas agreed to speak with immigration agent Jose Ovalle Jr., from the office of investigations in Harlingen. That very day she had been arrested, along with her boyfriend Juan Cisneros (also known as Juan Carlos Don Juan) at a mall in McAllen, Texas.

Erica agreed to speak to the immigration agent without the benefit of an attorney, because she wanted him to know that she was not guilty of the charge that she had helped transport Alexis, an undocumented 3-year-old from Honduras.

"I did not have anything to do with smuggling the boy. I was only accompanying him [Juan Cisneros]."

Erica also mentioned that she did not know Víctor Rodríguez very well, but that she had seen him on a few occasions, when he had come by the house in search of her boyfriend.

"He is an old man with a beard and a moustache," Erica recalled.

But then Erica contradicted the account of investigators. She stated that the boy, Alexis, was already in the car when her boyfriend Juan went to pick her up at the welfare office. According to the statements made by Juan and two agents, Erica had gone with Juan to Víctor Rodríguez's home to pick up the little boy.

At the federal court in McAllen, Texas, the judge was not fully convinced by Erica's statement. Erica was charged with possible conspiracy to transport an undocumented immigrant, with the expectation of receiving financial remuneration. But she was not charged with endangering the life of the child.

Despite the statements she made immediately after her arrest—statements that put her defense attorney in a very sticky situation—Erica was freed on bail six days later; someone had to take care of her baby. The judge did place one condition on her release, however: Erica had to find a job and keep it. But Erica didn't want to change the plans she had made to go back to school; her life was already complicated enough . . . and now this. In any event, she was lucky: she was the only one of all those arrested who had been released on bail.

Juan Cisneros, meanwhile, would follow all the legal proceedings from jail. His life with Erica was now over.

Juan, only 22 years old, spent fourteen months in jail before pleading guilty to harboring and illegally transporting a 3-year-old child from Honduras. Neither Juan nor Erica was charged with participating in the operation that caused the death of nineteen immigrants. And that was very good news for them. On July 19, 2004, a federal judge in Houston sentenced Juan, but only to the fourteen months he had already served in jail. Finally, Juan was free.

Freddy Giovanni García Tobar had many names. Karla, his girlfriend, just called him "Freddy." Tyrone Williams, the driver with whom Freddy and Abel Flores had supposedly negotiated the use of the truck for transporting undocumented immigrants, called him "Joe." But to the authorities on the hunt for him, the legal documents identified him as "Alfredo García."

García remained at large for almost a year. Of the fourteen individuals named by the U.S. authorities for their presumed involvement in the death of the immigrants in Victoria, Texas, Freddy was the last to be taken into custody. Ten had already been arrested or placed under surveillance in the U.S., and the other four had been arrested in Mexico.

García knew that they were looking for him, but one day he committed a foolish mistake. On Monday, April 26, 2004, while driving in Harlingen, Texas, the police stopped him for a traffic violation. Why had he remained in south Texas? Why hadn't he returned to Guatemala, his home country? How had he survived financially during the year he was on the run?

The minute they arrested him, the police instantly knew who had fallen into their laps. They had Freddy's Social Security number, his birth date (September 29, 1979), and the address of the house in Harlingen where he had lived until he'd gone on the lam. That same Monday, he was denied bail and on judge's orders he was transferred to Houston. There, a month and a half later, he would face several criminal charges that could put him behind bars for the rest of his life.

The prosecution finally had everyone they wanted.

Monday, June 14, 2004, was the day Karla Chavez's trial was scheduled to begin. Everything was ready. The immigrants who had survived the tragedy in Victoria had returned to Houston to testify against her and the other defendants. Nineteen of them were willing to testify in court that they had paid Karla to help them make the trip to Houston, Texas, and that she had been the person responsible for getting them into the truck container. In exchange, they had received work permits and Social Security numbers. As the result of the tragedy, they were now legal residents of the United States, albeit temporarily.

The news media was ready, as well. All the reporters knew that the trial could take weeks, maybe even months, of a prickly, humid summer in Houston, battered by squalls and floods. That morning, reporters from a number of English- and Spanish-language news organizations gathered in front of the court of U.S. District Judge Vanessa Gilmore. But nobody thought that anything transcendent would take place that morning.

It would be a long, drawn-out process. First, a jury had to be selected, and then they would all have to sit for hours and hours, day after day, listening to the details of the case: the evidence against Karla, the arguments presented by the defense, and then, much later on, the jury's verdict.

But none of that turned out to be necessary. In a surprising turn of events, Karla Chavez declared to the judge that she was guilty of having participated in an immigrant trafficking operation that ended the lives of nineteen people. It all happened so quickly that many reporters weren't even in the courtroom when she made her statement.

Karla, who has short, straight black hair, entered the courtroom, handcuffed and wearing her green prisoner's uniform. She was accompanied by her lawyer, John LaGrappe, who informed the judge that Chavez would be pleading guilty to the first of the fifty-six charges against her.

Things, however, did not appear to be quite so clear to Chavez herself. Instead of acknowledging her guilt, Chavez made the following statement to the judge.

"A lot of people died; I didn't kill them." She spoke in Spanish; her words were immediately translated by an interpreter.

That, without a doubt, was no confession of guilt. Judge Gilmore once again asked Chavez if she wanted to plead guilty to the first charge. Fearing that their agreement with Daniel Rodríguez, the prosecutor for the case, would disappear into thin air thanks to what she had just said, Karla's lawyer stepped in and told the judge that although she never had any intention to kill the immigrants, she did plead guilty to having caused their deaths.

LaGrappe's explanation did not satisfy the judge. Two more times, the judge asked Chavez, through the court interpreter, if she understood that she was pleading guilty to the deaths of nineteen immigrants. Following a moment of tension, Judge Gilmore accepted Chavez's guilty plea.

Yet there still remained a doubt in everyone's mind as to whether Karla really considered herself responsible for what had happened. Her statement of "I didn't kill them" is clear as a bell. It does not leave room for interpretation.

The other item left very clear that day in the courtroom was that Karla's lawyer had reached an agreement with U.S. Attorney Daniel Rodríguez and his team, who would dismiss the other fifty-five charges against Chavez and recommend a lighter sentence in exchange for Karla's testimony against the other defendants. Very possibly, this agreement might keep Karla, then 26 years old, from spending the rest of her life in jail.

Judge Gilmore set Karla's sentencing for spring of 2005.

Without a doubt, Karla's guilty plea was a tremendous coup for prosecutor Dan Rodríguez. In addition to locating and arresting all the people involved in the case, he had also managed to get a guilty plea out of the supposed leader of the fatal operation.

"In contemporary history, we haven't seen anything worse, so this is a significant guilty plea. We hope she will be capable of providing assistance to find out what other individuals were involved and to ensure that the individuals who have already been indicted are found guilty, and that everything that was done is disclosed," Rodríguez told the *New York Times.*[1]

Rodríguez's statements were a direct violation of Judge Gilmore's gag order, which prohibited all participants in the case from sharing their opinions with the news media. The judge reprimanded Rodríguez in public and threatened to take the case to another city. Even so, there was no denying that Rodríguez and his team had achieved a major legal victory.

Just like Karla Chavez, Abelardo Flores, Fatima Holloway, and Juan Carlos decided to plead guilty to some of the charges against them. And in exchange for lighter sentences, they might testify against Karla. Especially Abelardo Flores. He was willing to state that he had worked for Karla and that, following her instructions, he had helped recruit driver Tyrone Williams to transport the immigrants in the truck container.

In addition, at least nineteen survivors of the tragedy were willing to testify that Karla, in some way, had been the transportation coordinator for the entire group traveling in that trailer. But none of that testimony would be necessary. Karla's

---

1. *The New York Times.* June 15, 2004. "Woman, 26, Pleads Guilty in Deadly Smuggling Case," by Kate Zernike.

attorney convinced her that statements made public by the Justice Department, in addition to her own videotaped confession, had all but destroyed her chances for an innocent verdict.

Up until then, the other defendants had been planning to plead innocent in trial, but the prosecutors applied the same strategy with them that they had applied with Karla. Part of Karla's deal for a lighter sentence was that she would testify against the other defendants who were still maintaining their innocence. The prosecution's strategy was to pit the coyotes against one another. And it worked. Instead of maintaining their code of silence, one by one they succumbed to the pressure and started talking, pointing fingers at their ex-colleagues in the hopes of receiving lighter sentences in exchange.

In the end, the coyotes ate each other alive.

8
LAST WORDS

*"In contemporary history, we haven't seen any-
thing worse."*[2]
—DAN RODRÍGUEZ, federal prosecutor,
following Karla Chavez's guilty plea.

For the moment, when I saw him, I accepted [his death]. But
then afterward, I couldn't. A war broke out inside of me.
Deep inside. Why did all of this have to end up the way it
did? Why did it all end so sadly? I try to be patient about it."

Carolina Acuña,
little Marco Antonio's mother

Carolina now regrets not having sought help for her domestic violence problems, and she hopes that her story will serve as an example to other women who are going through similar situations.

2. *The New York Times.* June 15, 2004.

"I want them to see what happened to me. There is no such thing as 'should have.' What you don't do at a given moment, what you don't communicate . . . after that, you can't do anything, you can't turn back the clock. I didn't speak out. Lots of people knew, lots of people didn't, but I never said a thing about it. I kept it all inside, and that is never a good thing. I never even told his family about it. I didn't tell them anything. I didn't go to see them. I should have said something, even if only to say to him, 'This isn't right. We need to get help.' But no. I didn't ask for help because I said, 'Now what? What if he does it again anyway?' And then he did it again. That's why I would never tell a woman to keep her mouth shut. It doesn't do you a bit of good."

When I spoke with Carolina at a Mexico City hotel in November of 2003, it had been several months since she had last been abused.

"It's been six months, and nobody's hit me. That's an awfully nice feeling. I paid a very high price for it, though, I paid a very high price not to ever get hit again. I paid a high price, but other people shouldn't wait the way I did. Don't wait, fight it, because life is so much better when you're not getting abused, or humiliated, or yelled at. When you're free."

Carolina still talks to her son Marco Antonio as if he were alive.

"Thank you for making me happy, sweetheart. Happy! You gave me so much joy: your smile, your scent . . . we were once together, and that's what makes me strong, sweetheart. Onward. If I learned anything from you, it was that. That while you were alive, you wanted to be someone, sweetheart. Wherever you are, my love, I feel you close to me, sweetheart. You are no longer with me, but I can feel you at my side. And I am happy, Marco. I still have hope. I have hope that God will have mercy and that He will let me see you again someday."

Speaking on behalf of the Villaseñor family, Salvador Villaseñor del Villar, cousin of José Antonio, Marco Antonio's father, responds to Carolina's claim that she was the victim of domestic violence at the hands of José Antonio. In a television studio he said, "We fully believe that accusations are to be made when people are still alive, when people have the chance to clarify and solve their conflicts. Beyond that, however, we cannot comment on an issue that involves the woman who, after all, was Marquito's mother, and she is a woman we respect."

Salvador, however, did have a bit more to say about whom he felt was legally responsible for the deaths of José Antonio and Marco Antonio.

"We are convinced that the responsibility is shared by various groups, by the mafias that smuggle immigrants," he stated on behalf of the family. "But the governments of the United States and Mexico are also to blame for this. Why? Because of the conditions that immigrants must endure in order to get to the United States—all the abuses and humiliations are atrocious, unacceptable violations of human rights. And frankly, because the persecution strategy—that of treating people as criminals for trying to keep body and soul together, for hoping to find a better way of life—seems wrong to me. We also must face the reality that we, as citizens, are partially to blame, as well. How? In the sense that we, the citizens of the U.S. and Mexico have been unable to demand that our authorities address this issue. They don't care about immigration. It's that simple. But we have not managed to use our influence to shape policies that might create an organized immigration plan. Without a real leader, there can only be anarchy, and when there is anarchy, obviously, we end up with the situation we find ourselves in today."

Eduardo Ibarrola,
Consul General of Mexico in Houston

"The real solution, of course, is to fight the traffickers of human lives so that we can create a safe border and establish the basis for a new legal framework between Mexico and the United States, one that allows for an orderly, well-organized immigration system.

"We see that the United States has a real need for labor from other countries and obviously, Mexico is the closest country. As the population of the United States continues to grow older, the country is going to need more and more young workers who can replace the people who are vacating those jobs—workers who will pay taxes, who will be able to support the pensions in the United States, and who will help the U.S. continue to be competitive internationally.

"The issue of migrant laborers is one of globalization's many phenomena. Hopefully it will give rise to a new legal framework and a new regulatory framework that will facilitate the orderly, organized flow of migrant workers.

"Unfortunately, I believe that the migrant [workers] of Mexico and many other countries are willing to do anything. They are willing to assume whatever risks they must in order to make more money that they can send to their families and loved ones back in their home countries. This is very sad and very alarming. The authorities in both countries should work together to prevent situations like the one in Victoria from happening again, because I do believe they are preventable."

Lastenia Pineda,
consul general of Honduras in Houston

"The terrible hardships in our countries drive people to search for a better way of life; that is the 'American Dream.' They have a lot to do [with that dream] because they are the ones who have become great business leaders. But it is the hardships in our countries [that are to blame for illegal immigration]."

Martín Berlanga,
Univision correspondent in Texas

"Some stories affect you as a journalist, and other stories affect you as a person. That's what h~ened to me in this case. Watching them take the body of the little boy's father out of the truck really drove things home. And then I saw the photographs of the little boy. I have children of my own, and just picturing myself in that situation . . . Few other stories have ever had that effect on me."

Enrique Ortega Cuate, survivor

"They chase me, and then people start fighting. As if I was running from the immigration agents, they chase me and then I see bodies in blood. I guess that's because afterwards, the photos of the bodies were filled with blood. That might be why [I dreamt that], I don't know."

Alberto Aranda Amaro, survivor

"When I came here, I said to them, 'I want to see my house finished. I don't want us to have any more problems with the neighbors because we don't have fences up.' That was always our dream: to live well. Ever since the first day I got here, I started sending money to my mother, and she would stretch the money as far as she could, and she would say to me, 'Listen, son, this is what I did.' I still can't say that I've achieved my goal but I'm getting there. As long as God keeps me on this earth and gives me strength, we're going to get there."

After the tragedy in Victoria, almost a year went by before Alberto Aranda Amaro could send his mother any more money. But finally, in March of the year 2004, he finally did. At his house in Mexico, Alberto built a small altar in honor of the Virgin Mary. "With the money you sent me, I'm going to buy some flowers for the virgin," his mother told him. He never found out what kind of flowers they were, but he was sure that it was "a very special bouquet."

"There are a lot of immigration agents around here," he said. "But things are easier for me now." Just like the other survivors from Victoria, Alberto was granted a Social Security number and a work permit so that he could stay in the U.S. until he was called to testify in the trial. The new immigration service wasn't what tormented Alberto now: it was the darkness of the night that plagued him. Alberto continues to dream about the bodies of the immigrants that died in the trailer. The immigrants, unfortunately, were not given any kind of psychological treatment to help them cope with this traumatic experience.

Alberto is not the only one with this problem. Once, at the office for immigration services in Houston, Alberto ran into Matías's aunt. Matías, the boy Alberto helped down from the trailer, wasn't there because he was working.

"Sometimes Matías wakes up screaming," Matías's aunt told Alberto. "Do you have those same nightmares?"

Alberto can at least rest assured in the knowledge that he did all he could to help look after Matías.

"I think maybe he respects me a little. He thinks that I saved him. He felt very alone, and I gave him strength. It was his first time coming to the United States."

"What would you do?" he once asked Enrique over the phone. He and Enrique were the ones who had broken the tail lights at the back of the trailer. "Do you think that [Karla and the other defendants] deserve the death penalty?"

"She's so young," Enrique answered, not responding directly to Alberto's question. "But it's like a religion with them; it's the only way they'll be satisfied."

"Not me," said Alberto. "If someone in my family had died, maybe I'd want them to get it. But killing someone isn't going to solve anything."

José Reyes Arellano Gaviles, survivor

"All I do is just ask God, who let us live, to never let anyone else go through what we did. It makes me so sad to think of how my brother-in-law and his two uncles died. They're with God now, in His holy kingdom."

Israel Rivera Sanchez, survivor

"I regretted it. Man, if I hadn't come, if my uncle hadn't come, none of this would be happening to us. This was my first time in the United States. I came because I was hoping to find work. But then, with this terrible tragedy that happened to us, sometimes it gets you thinking, and I just start getting so sad about everything we lived through in that trailer. It was so awful what happened, and I don't wish it on anyone."

Jim Kolbe, Republican congressman from Arizona, made these statements to the Gannett News Service regarding the deaths of immigrants on the U.S.-Mexico border

"I wish we could make people wake up to see what's happening. People [in Washington] see it as a distant event, like reading about an earthquake in Turkey in the newspaper. They feel bad about it, then turn the page."[3]

Karla Chavez, after pleading guilty, on June 15, 2004, to conspiring to transport the immigrants who died in the trailer

"I didn't kill them."[4]

3. Representative Jim Kolbe R-Arizona to the Gannett News Service (June 11, 2003)
4. Karla Chavez, June 15, 2004.

ACKNOWLEDGMENTS

The majority of this book is based on the testimony of various survivors of the tragedy in Victoria, Texas. This book would never have been possible had it not been for the conversations I recorded with Enrique Ortega, Alberto Aranda Amaro, José Reyes Arellano, and Israel Rivera Sanchez. These four men were kind enough to allow me to publish the interviews that I had originally taped with them for a television program. Later on, I reconnected with Enrique, Alberto, and José to expand on the information they gave me and to clarify some of the details and issues that had come up during our long conversations in Houston and Victoria, Texas.

Thank you, Enrique. Thank you, Alberto. Thank you, José. Thank you, Israel. Without you, an important part of this story would have disappeared into thin air.

Carolina Acuña, little Marco Antonio's mother, spent hours with us in Mexico City, telling us the story of how her son ended up going to the United States and of how she thinks of him every day. Thank you for your courage, Carolina.

Salvador Villaseñor del Villar, cousin of José Antonio Villaseñor (Marco Antonio's father), spoke frankly and openly on behalf of the Villaseñor family. Thanks to our conversations with him, we came out with a more balanced perspective on the events that led up to the deaths of José Antonio and his son Marco Antonio. Salvador valiantly fought his own pain to give us a clear picture of the daily drama endured by millions of Mexicans who dream of going north, and his reflections were very helpful in putting the issue into perspective for us. Thank you for that.

The genesis of this book, I should note, was the investigative reporting carried out by the courageous and talented team of journalists from the news department at Univision. Marilyn Strauss and Lourdes Torres were the executive producers of the program *Viaje a la Muerte (The Road to Death)* that originally aired in November of 2003. Evelyn Pereiro, Angel Matos, Porfirio Patiño, and Martin Guzmán are just a few of the many people who contributed to this project.

All my editorial projects have been extremely fortunate, for they have always received the generous cooperation and support of Univision, and this project is no different. In particular the efforts of Ray Rodríguez, Frank Pirozzi, and the vice president for news, Sylvia Rosabal, were crucial in helping me transform this very special news program into a book.

I would also like to thank Henry García Castillo, the sheriff of Victoria County, Texas; nurse Gilda Miller of Citizens Medical Center; and María Ortega, the sister of one of the immigrants, for the statements they offered Marilyn Strauss, special events producer for Univision, and which I include in the book.

The interviews I conducted with Lastenia Pineda, the Honduran consul, and Eduardo Ibarrola, the Mexican consul, both in Houston, provided me with some very vital background and context for this book; our conversations were crucial, for they gave me a better understanding of the reasons and circumstances that led up to this terrible incident. Carolina Zaragoza, the Mexican consul for immigrant protection, was a marvelous and generous source of support for many of the survivors, defending and protecting them far beyond her diplomatic obligations. She was also a conscientious, respectful liaison for many of the people I interviewed. Carolina knows more about this case than anyone. And she was kind enough to share much of her knowledge with me. Thank you so much, Carolina.

The Devil's Highway, Luis Alberto Urrea's magnificent book about the death of fourteen immigrants in the Arizona desert, convinced me that this was the kind of story that had to be told: if there are no accusations, if there is no sense of urgency, nothing will ever change in that immense cemetery we call the U.S.-Mexican border, a cemetery that is only growing larger with each passing day.

I wish I had never had to write this book in the first place. But the story of the nineteen men who died and the others who survived that night in Victoria is one that must be told in the hopes that history will not repeat itself in this way again.

Note: A portion of the proceeds from this book will be donated to immigrant-advocacy organizations both in and out of the United States. For more information on these organizations and on how to get in touch with them, please visit the web site www.jorgeramos.com.